Artificial Intelligence Accelerates Human Learning

Katashi Nagao

Artificial Intelligence Accelerates Human Learning

Discussion Data Analytics

Springer

Katashi Nagao
Nagoya University
Nagoya, Japan

ISBN 978-981-13-6174-6   ISBN 978-981-13-6175-3   (eBook)
https://doi.org/10.1007/978-981-13-6175-3

Library of Congress Control Number: 2018968397

© Springer Nature Singapore Pte Ltd. 2019
This work is subject to copyright. All rights are reserved by the Publisher, whether the whole or part
of the material is concerned, specifically the rights of translation, reprinting, reuse of illustrations,
recitation, broadcasting, reproduction on microfilms or in any other physical way, and transmission
or information storage and retrieval, electronic adaptation, computer software, or by similar or dissimilar
methodology now known or hereafter developed.
The use of general descriptive names, registered names, trademarks, service marks, etc. in this
publication does not imply, even in the absence of a specific statement, that such names are exempt from
the relevant protective laws and regulations and therefore free for general use.
The publisher, the authors and the editors are safe to assume that the advice and information in this
book are believed to be true and accurate at the date of publication. Neither the publisher nor the
authors or the editors give a warranty, express or implied, with respect to the material contained herein or
for any errors or omissions that may have been made. The publisher remains neutral with regard to
jurisdictional claims in published maps and institutional affiliations.

This Springer imprint is published by the registered company Springer Nature Singapore Pte Ltd.
The registered company address is: 152 Beach Road, #21-01/04 Gateway East, Singapore 189721,
Singapore

Preface

Discussion is something that often occurs in everyday conversation, for example, sharing your opinion with others and answering questions. However, it is slightly more sophisticated than just chat. People who are good at communicating are generally good at discussion.

In the case of communication ability in general, you might think of someone that considers the feeling of their conversation partner, or thinks about what is appropriate conversation for the current environment, or maybe even that they are more skilled at making small talk. Although I even believe that these are valid, the most essential skill to acquire is discussion ability. In addition, I would like to explain about three very important technical concepts in this book.

The first is data analytics, which is a scientific system for gathering objective facts (data) in large quantities, analyzing data probabilistically and statistically, and making new discoveries. The second is natural language processing, a technique for analyzing and generating human language. The third type is gamification, which is a methodology that introduces game elements (such as scoring points and competition, to gradually increase the skill level of the target by giving a sense of accomplishment, etc.) for daily activities.

These three concepts play an important role in efficiently improving discussion ability. Data analytics is used in the objective analysis of human behavior, and natural language processing is used to extract features of human behavior, especially to extract and utilize features contained in words. One should think about training for discussion ability in the same way as training for improvement in sports. In other words, we need self-analysis and practice to overcome weaknesses and improve strengths.

However, different from sports, discussion is an intellectual activity, and those data are composed mainly of words. Although the technology concerning the analysis of words is considerably advanced, it is still very difficult to analyze its semantic contents with high accuracy. So, I will propose a creative idea. That is to attach attributes to people's remarks. We think about what we want to discuss next, and decide if it is related to the previous speaker's speech or not, then we categorize the speech. Of course, if you follow conversation properly as a human, you can understand without the need for categorization. However, it is quite difficult for a machine to judge the relatedness of remarks in the same manner. Interestingly, if you were to ask most people if they were aware in what way their utterances were related to the flow of the conversation, you would find that in most cases they are not completely aware. To tell the truth, it is good for both machines and human beings to listen to other people's speech and think about the connections.

We now think about how to apply data analytics to a discussion, where the simplest analysis is to examine the number of remarks produced by each participant. After collecting data for around a year, one of the first things we found was that people with a higher number of utterances were found to have a higher level of communication ability in comparison to those who produced less utterances. Of course, there are some people that make many utterances with little content. However, we have established practical policies for our meeting system in our lab that encourage thoroughly listening to the speech of others before talking during their turn and to produce utterances that are related to the previous speech or if it is not related to ask clear questions.

Next, we examine the relationship between a remark and the preceding utterance. This makes it easier for analysis by having metadata on speech. What I learned from the analysis is that people who are good at discussions tend to make related remarks when someone speaks related to their remarks. In other words, people who are capable of making related statements consecutively will have high communication skills.

By using data analytics, you can explain such things with a clear basis.

On the other hand, what is gamification? Applying game-like elements to everyday activities is not an entirely new concept. We all are familiar with the teacher praising the first student to get a question right in front of the class.

It is an important requirement for games to reward achievement of tasks, allow competition with multiple people, and show results in an easy-to-understand manner before the player grows bored of the game, but this also can apply to daily activities where it isn't rare to gauge motivation.

Nevertheless, why try to systemize games with a name like gamification? To that end, information technology (IT) is also heavily involved. Data are important here as well. In short, IT is a mechanism that processes digitized data to suit the circumstances of human beings, so gaming can also actively utilize the data to expand its effect. For example, a small survey can now be easily filled out via smartphone with the results being quickly summarized for everyone present to view. If you happen to have past data from a similar survey, you can quickly compare it and carry out statistical processing. In this way, IT is providing humans with new opportunities to think. The fact that we are able to collect and quickly process information for human understanding will change our definition of what is possible.

So, I would like to try to make discussion a game. Some might think that there already exists debate which is like discussion made game-like. Debates have clear rules and frameworks to decide the outcome. Given a problem, such as "whether the Japanese constitution should be revised", arguing and arguing against and against, we will compete to determine which claim is more convincing. We can imagine that

Preface

debate will be good training for discussion. However, it is far from the communication which we usually do. We will express our opinion based on our thinking and experience. It is constructive to train people thinking about how to express such opinions effectively to the opponent. Debate is not appropriate for it.

Then, what should we do?

Rules like debate are not necessary for training to improve discussion. However, there are some restrictions. In order to introduce gamification, at least you need to do the following. First, you have the speaker record their name when they speak and whether their remarks are related to previous statements. This also makes it easy to use IT, so that you can enter it automatically for the former speaker. Next, when someone's remarks are over, we will evaluate that statement. This evaluation is a little difficult, but here it is easy to think that it is whether you were convinced by the remark or not by entering yes or no. In fact, you may have to enter somewhat more complicated things, but that is when the stage in the game has advanced.

The way to train the discussion ability using IT like this is very effective. We have been doing research on this for over 10 years. This book is written to share the research results with readers. In addition, I will explain data analytics and gamification in order to understand the research contents. Of course, I would like to touch on artificial intelligence which is the cutting edge of IT, especially machine learning and data mining.

It would be an unexpected pleasure if readers would someday become good at discussions and communication by utilizing our research.

I greatly appreciate the assistance provided by people who contributed to this book. The staff and students of Nagao laboratory of Nagoya University, Shigeki Ohira, Shigeki Matsubara, Katsuhiko Kaji, Daisuke Yamamoto, Takahiro Tsuchida, Kentaro Ishitoya, Hironori Tomobe, Kei Inoue, Kosuke Kawanishi, Naoya Kobayashi, Naoya Morita, Saya Sugiura, Kosuke Okamoto, Ryo Takeshima, Yuki Umezawa, Shimeng Peng, Ryoma Senda, and Yusuke Asai, whom I have worked with and have kindly supported me during the development of the prototype systems introduced in this book. Kazutaka Kurihara, Professor of Tsuda University, provided me with much helpful advice to develop the ideas described in this book.

Also, I thank Miku Suganuma, Miyuki Saito, and William Samuel Anderson who gave me helpful advice on the wording used in the book.

The work described in Chap. 5 was supported by the Real-World Data Circulation Leaders' Graduate Program of Nagoya University. I thank Kazuya Takeda, program coordinator, and Mehrdad Panahpour Tehrani and Jovilyn Therese B. Fajardo, designated professors of the program, who are co-authors of the journal paper based on Chap. 5.

Furthermore, I am indebted to my editor, Mio Sugino, who has been a constant mentor during the writing process.

Finally, I am particularly grateful to my wife, Kazue Nagao, and my daughter, Saki Nagao, without whom I could not have undertaken this effort.

Nagoya, Japan

Katashi Nagao

Contents

1 Artificial Intelligence in Education 1
  1.1 AI and Education 1
  1.2 e-Learning 4
  1.3 Intelligent Tutoring System 7
  1.4 Learning Analytics 9
  1.5 Machine Learning Accelerates Human Learning 11
  1.6 Deep Learning Approaches 13
  References 16

2 Discussion Data Analytics 19
  2.1 Structure of Discussion 20
  2.2 Discussion Mining System 21
  2.3 Structuring Discussion 23
  2.4 Summarization of Discussion 25
  2.5 Task Discovery from Discussion 27
  2.6 Active Learning for Improving Supervised Learning 33
  2.7 Natural Language Processing for Deep Understanding
     of Discussion 37
  2.8 Natural Language Processing for Discussion Structure
     Analysis 41
  2.9 Correctness of Discussion Structure 46
  2.10 Structuring Discussion with Pointing Information 51
  References 55

3 Creative Meeting Support 57
  3.1 Meeting Analytics 58
  3.2 Machine Learning for Structured Meeting Content 63
  3.3 Post-meeting Assistance to Support Creative Activities 64

3.4	Evaluation of Creativity		73
3.5	Future of Creative Meeting		74
References			75

4 Discussion Skills Evaluation and Training ... 77
- 4.1 Evaluation of Speaking Ability ... 78
- 4.2 Feedback of Evaluation Results ... 83
- 4.3 Evaluation of Listening Ability ... 88
- 4.4 Discussion Skills Training ... 93
- 4.5 Gamification for Maintaining Motivation to Raise Discussion Abilities ... 94
- References ... 103

5 Smart Learning Environments ... 105
- 5.1 Environments for Evidence-Based Education ... 106
- 5.2 Related Work and Motivation ... 106
  - 5.2.1 Discussion Evaluation ... 107
  - 5.2.2 Presentation Evaluation ... 109
  - 5.2.3 Motivation ... 109
- 5.3 Leaders' Saloon: A New Physical–Digital Learning Environment ... 110
  - 5.3.1 Discussion Table ... 110
  - 5.3.2 Digital Poster Panel ... 110
  - 5.3.3 Interactive Wall-Sized Whiteboard ... 111
- 5.4 Importing Discussion Mining System into Leaders Saloon ... 112
  - 5.4.1 Discussion Visualizer ... 112
  - 5.4.2 Discussion Reminder ... 114
  - 5.4.3 Employing Machine Learning Techniques ... 116
- 5.5 Digital Poster Presentation System ... 116
  - 5.5.1 Digital Posters Versus Regular Posters ... 116
  - 5.5.2 Authoring Digital Posters ... 117
  - 5.5.3 Data Acquisition from Interactions with Digital Posters ... 119
- 5.6 Skill Evaluation Methods ... 120
  - 5.6.1 Discussion Skills ... 121
  - 5.6.2 Presentation Skills ... 122
- 5.7 Future Features of Smart Learning Environments ... 125
  - 5.7.1 Psychophysiology-Based Activity Evaluation ... 127
  - 5.7.2 Virtual Reality Presentation Training System ... 130
- References ... 134

6 Symbiosis between Humans and Artificial Intelligence ... 135
- 6.1 Augmentation of Human Ability by AI ... 136
- 6.2 Intelligent Agents ... 139

6.3	Singularity	143
6.4	Human–AI Symbiosis	144
6.5	Agents Embedded in Ordinary Things	145
6.6	Artificial Intelligence Accelerates Human Learning	147
References		151

Chapter 1
Artificial Intelligence in Education

Abstract This book explains how human learning is promoted by applying artificial intelligence to education. Before that, let's first look back on how information technology including artificial intelligence contributed to education. Various technologies have been developed to make it easier for learners to learn and to create an environment where teachers can more easily teach. An example of this is called e-learning or intelligent tutoring systems (ITS). e-Learning is an educational system using online media and has developed together with web technology. ITS was developed using a rule-based system which is an initial result of artificial intelligence. In the process, user models for learners called learner models and educational contents have been improved. As an application of data science, technology called learning analytics was developed. This is a technique for statistically analyzing learner's historical data obtained by e-learning, etc. and discovering the characteristics of the learner. This will contribute to personalized learning that adapts the educational system to the learner's characteristics. Furthermore, the development of learning analytics will clarify the concept of evidence-based education. As with medical care, we should construct a feedback loop that educates in accordance with data-based analysis and the learning strategies obtained from it, and improves if there are problems. Machine learning, which is an important achievement of recent artificial intelligence, is used for data analysis at this time. In addition, we will use a method that lets machines do the feature extraction from data called deep learning. In this chapter, I will touch them in detail.

Keywords Intelligence amplification · e-Learning · Intelligent tutoring system · Learning analytics · Evidence-based education · Deep learning

1.1 AI and Education

Research on an educational support system using artificial intelligence has been exploring important questions such as how to support learning and problem-solving which can be considered as the foundation of human intelligence and how to communicate with people to realize it (Wenger 1987).

© Springer Nature Singapore Pte Ltd. 2019
K. Nagao, *Artificial Intelligence Accelerates Human Learning*,
https://doi.org/10.1007/978-981-13-6175-3_1

It is also a research area that has led and impacted critical issues such as natural language understanding and data mining as well as essential technologies for educational support as well as artificial intelligence research.

While research on artificial intelligence focuses on the realization of machine intelligence, research on educational support is of primary interest in raising human intelligence through human learning and support for problem-solving.

Approaches to such intelligence are often referred to as intelligence augmentation or intelligence amplification (IA).

However, the approaches are consistent in that they explore human intelligence even if they are different, and research on educational support systems has been trying to elucidate the way of learning, problem-solving, and communication from the viewpoint of IA.

The goal of such research on educational support systems is to realize intellectual education/learning support for learners.

What is "intellectual" here? Although the answer is not fixed, it has been aimed at realizing intellectual support from almost two perspectives so far.

One is to adaptively adapt to individual learners, that is, adaptation according to the characteristics of each learner from the viewpoint of "teaching".

The other is to provide appropriate learning environments and tools for learners from the viewpoint of "learning", to build a foothold for problem-solving and learning, called scaffolding.

In order to realize these supports, the design and development methodologies of systems and element technologies have been explored while utilizing theories and knowledge in related fields such as cognitive science and pedagogy.

The origin of such inquiry dates back to the study of intelligent computer-assisted instruction from the 1970s to the 1980s.

From that time on, the composition of the system necessary for realizing intellectual educational support with the following four factors (1) teaching knowledge expressing how to teach, (2) teaching material knowledge showing teaching knowledge, (3) knowledge state and understanding state of the learner, and (4) the user interface to realize bidirectional dialogue (Carbonell 1970) has been taken into consideration, and these elements have directed the research and development of education and learning support technologies up to today.

In particular, it has been found that it is indispensable for the advancement of intellectual support by the system to represent the various knowledge necessary for support structurally.

This point is consistent with the methodology of expert system development in knowledge engineering (Clancey 1987) and has had a very significant impact on subsequent educational support system development.

Especially in the educational support system research, emphasis shifted from educational orientation to learning oriented since the 1990s, and the paradigm of research has shifted to a learner-centered system (Norman 1996).

Progresses in information and communication technology pushed this paradigm shift.

1.1 AI and Education

Today, based on these research backgrounds, a research methodology that systematically constructs support methods by clarifying the way of learning support by modeling learning and its support as an information processing process and expressing it on the system has been promoted (Kashihara 2015).

Research based on this methodology is called "learning informatics".

In realizing intellectual educational support, the ability of the system to properly understand learners is essential, even though domains, tasks, and learning contexts to be supported diversify in the future, its importance remains unchanged.

However, even if the human teacher does not fully understand the learner, it is possible to provide accurate support through interactions, so it is not necessary to aim for building a cognitively sophisticated learning model.

It is important to generate learner model information inside the system so as to contribute to the effective execution of educational support or design interaction to compensate for imperfections in learner model information.

On the other hand, it is often difficult to generate adequate learner model information, such as mastering higher order skills such as problem-solving, learning, and creation in unstructured domains.

In response to such problems, a framework is proposed to disclose learner model information generated by the system and to make learners themselves discover inadequate/inappropriate points of the model so that a whole system including learners as a part of the system can be modeled (Bull and Kay 2007).

Such a model is called an Open Learner Model. This approach also leads to encouraging learner's introspection by scrutinizing information on the learner model, and high learning effect is expected.

Also, even if errors and knowledge states found from learners' inputs are generated as learner model information, learners do not necessarily recognize them.

For such deviations between model expressions and learner's recognition, methods to promote self-recognition for errors by visualizing errors have been proposed.

These can be thought of as an attempt to effectively supplement information lacking as a learner model expression by interaction.

Furthermore, in the e-learning environment, enormous learning log data from the learner group is accumulated in learning management system (LMS) and e-portfolios, and this large-scale data can be analyzed. Then, it is possible to discover learning behavior patterns and new attributes that characterize learners that were not supposed beforehand.

This activity is called learning analytics and is an attempt to convert the amount of data into quality data and to find a new learner modeling technique (Ochoa et al. 2014).

1.2 e-Learning

Educational support using computer networks has been practiced for quite a long time, but e-learning, which is a form of educational training utilizing the Internet since the latter half of the 1990s, has been developed and is now widespread.

e-Learning in a narrow sense refers to an embodiment of education and training using a computer network.

For example, in asynchronous learning by web-based training (WBT), the learner learns using the teaching materials stored in the WBT server using a web browser, and the instructor grasps the progress of the learner with the learning log accumulated in the server.

In synchronous learning, simultaneous distribution of lectures and acceptance of questions from learners are realized using a video conference system.

It can be thought that e-learning in a broad sense includes not only implementation of education and training but also IT support for operation management and performance evaluation.

For example, not only WBT and synchronous learning but also learning progress, guidance history, schedule, used assets, etc. of a complex curriculum combining multiple educational forms such as group education, practical training, seminars, etc. are applied to a learning management system (LMS). Supporting the smooth implementation of the curriculum is also a major embodiment of e-learning.

Additionally, in thinking about things in phases for education and training in practice, we need to realize that there is more to be concerned with than just students using the systems; we have a beginning planning phase where we need to think about plans for what and how we will teach and a post-evaluation phase where we need to evaluate our measurements concerning educational effectiveness.

In this way, it is part of e-learning to manage the achievement goal of the curriculum and the achievement progress level of individual learners and to support feedback for the implementation of the next curriculum.

The main information in e-learning is related to content/learner/learning system/learning history. Standardization of data models and representation formats (bindings) on these are underway (Nakabayashi 2002).

In addition, interfaces for linking learning support systems and tools are becoming more standardized. A prime example of this would be the content standard shareable content object reference model (SCORM) which has been decided on by the American standardization organization the advanced distributed learning initiative (ADL).

The SCORM content consists of a course structure loaded into the WBT server, sequencing rules and metadata attached to the course structure, and shareable content object (SCO) executed on the Web client.

The course is described by a hierarchical teaching material structure (corresponding to the table of contents of the textbook), and the terminal page of the hierarchy corresponds one-to-one with the SCO.

1.2 e-Learning

The LMS sequentially selects pages to be presented to the learner from the course structure and displays the corresponding SCO on the screen of the Web client.

SCO is Web content composed of HTML, JavaScript, various plug-ins, etc., and includes explanation pages of exercise materials, exercise problem pages, simulation pages, and so on.

By describing the sequencing rules, it is possible to create learner adaptive content that dynamically selects the SCO that the LMS presents to the learner according to the solution situation of the exercise problem.

The standard specifies the data model of the course structure and the binding to XML and the API and data model (including learner ID, practical problem acquisition situation, learning time, etc.) of SCO/LMS communication.

According to the SCORM standard, the same teaching material can be executed by several different LMS (interoperability), and SCO can be used in combination with several different course structures (reusability).

Among the standards of content, there is a question and test interoperability (QTI) standard developed by IMS (IMS Global Learning Consortium, Inc.) as a standard specialized for testing (Nakabayashi 2008).

The QTI standard is roughly divided into standards relating to test content and evaluation results.

In the test content standards, a hierarchical structure of the entire exam question and a description method of the content of each question are stipulated.

A question can be categorized in many ways, such as in a written format, as multiple choices, true or false questions, by presentation style, and result evaluation.

In the hierarchical structure part, it is possible to designate the method of ordering the question, the method of feedback to the learner, the method of summarizing the result, and so on.

In the standard relating to the evaluation result, a format for recording the results, the number of trials, the required time, etc. of each question is stipulated.

That is, the entire QTI standard includes not only static descriptions such as question questions of questions, order of correct answers, and problems, but also descriptions of dynamic actions at the time of execution, such as correctness judgment method and resulting aggregation method.

In addition, as a content-related standard, a LOM, which is a metadata standard for describing attributes of LO (learning object), is formulated by the IMS LTSC (Learning Technology Standard Committee).

Here, LO refers to all digital/non-digital resources used for educational training.

LOM is indexed information for searching and reusing these resources.

The index information is composed of general information such as title and explanation of contents, education field, subject learner, degree of difficulty, and other education-related information, intellectual property right information, information showing relation with other LOs.

By constructing a repository of LO using LOM, it becomes possible to classify and extract LO according to educational purposes and to organize LO according to curriculum.

As a learner, the information standard IMS has formulated the Learner Information Package (LIP). LIP is a standard for describing attributes of learners and includes security information with identification information such as name and ID, learning purpose information, possession qualification information, learning history information, competency information, and password as main items.

According to the LIP standard, a format for exchanging student information between systems is standardized.

ISO/IEC 20006 Information Model for Competency that Japan proposed from ISO/IEC JTC 1 SC 36 exists as a standard for describing learning systems and skills learned by learners.

The standard specifies a framework for describing the learner's skills and a method for expressing standards for expressing skills.

As a standard for describing various learning methods (learning teaching strategies), the learning design (LD) standard of IMS standardized based on educational modeling language (EML) developed at the Netherlands Open University. EML and LD describe various learning teaching strategies. These strategies assume that "in learning, people belonging to a specific group, people with a specific role are involved in learning activities using an environment consisting of appropriate resources and services."

Descriptions are very diverse from self-taught teaching strategies that perform strict learning control to problem-solving cooperative learning by multiple learners.

Regarding the learning history standards, attempts were made to formulate standards related to portfolios, but they have not spread to widespread use.

In recent years, due to popularization of learning using mobile terminals and the like, standards have been proposed that simply define an interface to transfer learning history between systems.

One of them is the Experience API (xAPI) formulated by ADL.

xAPI is a standard for collecting the history of learning activities using mobile terminals and electronic devices such as electronic textbooks.

The target learning activity ranges from very micro-actions such as browsing web pages and page scrolling of electronic textbooks to a range of relatively large grain sizes such as evaluation results of exercise problems handled in SCORM and QTI.

We will explain the utility of technical standardization and its future direction.

The basic idea of technical standardization is to divide the whole system into parts (modules) and standardize the interface between the modules which enables the exchange of modules without impairing the function of the whole system.

This makes it possible to replace high-performance modules and add modules with new functions, which allows us to autonomously decentralize the evolution and value improvement of the whole system which is the essence of the effectiveness of technical standardization (Nakabayashi 2010).

Users can freely "exchange" and select items of favorite things such as performance and price from the products of several suppliers.

By giving users freedom of choice, there is a necessity to provide high-quality, low-priced products to suppliers, and competition will accelerate the improvement of value.

In addition, a module developer can easily develop and add modules that have new functions and convenience without knowing the detailed structure of other interlocked modules in the system.

In order to further promote such evolution, it is desirable that the standard be as simple and robust as possible.

The success of the Internet, especially the Web, is a good example of the drastic increase in value brought about by such simple and robust standards.

In the field of e-learning, the module that brings about such value improvement was originally a system, and after that it was contents.

Although it seems that their importance will not be lost in the future, on the other hand, the value of learner information and learning history information will increase more in the near future.

By standards related to learning history information such as xAPI, development of a learning analytics module for learning portfolio (e-portfolio) and large-scale learning history information is promoted. It is expected that a learning support function will emerge with higher added value than using these results.

1.3 Intelligent Tutoring System

An intelligent education system or intelligent tutoring system (ITS) supports learning through adaptive and interactive knowledge exchanges between a teacher's computer and learner using artificial intelligence technology (Wenger 1987; Woolf 2009, Nkambou et al. 2010). It consists of four modules: a teaching material module, a learner model module, a tutoring strategy module, and an interface module.

The teaching material module not only stores the knowledge that is the object of education but also has functions such as evaluating the learner's answer and providing information for problem-solving/question generation/explanation.

The learner model module observes the behaviors of learners, constructs models of learning situations such as emotions and motivation, understanding of learner's teaching materials, and provides necessary information to realize adaptive education.

Based on that information, the tutoring strategy module will decide and implement hints appropriate to the state of the learner and actions such as selection and explanation of problems.

The interface module then implements a multimodal interaction between the learner and the system.

The characteristic of ITS lies in modeling.

It has models of teaching materials, models of understanding by learners, and models of educational behavior, and support learners' learning while utilizing them dynamically.

In particular, since the student model (learner model) is the main source of adaptive behavior of the system, much research has been done. However, it is difficult to learn from a model of human understanding with a small number of observations. One could say that developing necessary and sufficient learning models is an eternal challenge for ITS.

A cognitive apprenticeship model (Collins et al. 1987) expressing a process to acquire cognitive skills is one of the influential models for acquiring and improving skills related to "learning".

This is an extension of the process of apprenticeship practiced by craftsmen with skills when developing disciples to cognitive skills, and the process is organized as follows:

1. Modeling: method of knowledge transfer by experts.
2. Coaching: presentation of hints and feedback by experts.
3. Scaffolding (foothold making): providing clues (scaffolding) according to learner's proficiency level.
4. Articulation (clarification/externalization): linguistic representation and externalization of thought by learner.
5. Reflection: review of learning process by learners.
6. Exploration (inquiry/practice): preparation of an environment where experts solve problems to learners.

In the early stages of cognitive skill learning, it is important to provide the student with the structure of the skill to be learned at an appropriate level of abstraction according to the level of the learner.

In order to realize this, it is desirable to understand that the structure of the skill to be learned is organized as a combination of a learning curve and an ontology, and the level visualization method by rubric, etc. is defined according to the type of skill.

In addition, as a method of presenting to learners, a method such as a concept map that structurizes relationships between concepts has been proposed to support a bird's-eye view of overall images.

Also in the early stages of cognitive skill learning, immediate feedback according to a learner's interaction is required to prevent wrong knowledge acquisition and a deadlock situation.

Specifically, there are issues such as assignment of tasks, division of problems, presentations of points of interest for orienting learning activities, provision of hints to learning activities, visualization of errors, and so on.

A method of adaptively recommending or dynamically generating candidates for the next learning activity according to the situation of progress may also be considered.

Increasing the slope of the development stage graph in the learning curve can be regarded as reaching a high skill level with a smaller number of learning times.

To provide such an appropriate "scaffolding", it is necessary for the system to grasp the skill level of a learner's cognitive skills and to control the support function provided by that level.

To realize this, it is necessary not only to pay attention to the learning results but to collect data reflecting the learning processes. We are utilizing rules like a fading technique which reduces support content based on the user's proficiency and an adaption technique based on learning style and learning method preferences to manage support functions.

It is one of the important learning activities that the learner himself/herself externally (lexically) represents the learning process at the developmental stage of learning of cognitive skills.

By doing so, the learner can objectively grasp his/her own state which is difficult to reproduce in reality.

In order to positively carry out such externalization activities, as an environment for externalization, a method of annotations such as an intention at the time of learning and an outline of learning activity and a method of drawing relations between concepts exist.

It is used in combination with supporting methods such as the externalization direction of the learning process.

It is not easy for students to notice the necessity of reflection (awareness) at the developmental stage.

On the other hand, excessive support that damages the learner's identity lowers the awareness to the learning process and learning outcomes, which makes it difficult to promote reflection.

In order to promote reflection, it is required that we attempt to "provide the environment in which the student himself/herself reflects as much as possible."

In addition to looking back on their activities, not only capturing changes in their performance but also approaches such as grasping improvement points through comparison with others are often performed.

"Learning how to learn" is more implicit than traditional "learning of knowledge," the period of learning is long, and the correct answer is not always unique.

For this reason, it is difficult to establish in advance an absolute criterion on how the proposed function contributed to the learner.

Specific solutions include an approach that extracts an ideal learning process such as a role model by utilizing social learning context by a large number of learners, and an approach that uses a similar learning process to personalize learning support functions. Also, an approach to evaluate the correspondence relationship between learning process and performance is conceivable. It is necessary to reduce the burden on the teacher and implement an interface that utilizes natural language.

In addition, how to present objective evaluation indexes is a major challenge for the future in utilizing ITS as an open learning environment.

1.4 Learning Analytics

In recent years, methods for visualizing, analyzing, and evaluating learning activities and their practical application are increasingly concerned.

Learning analytics is a method of measuring, collecting, analyzing, and reporting evidence-based data on learning and context in order to optimize and improve the learning process and learning environment. It is based on online education such as massive open online courses (MOOCs) (Watanabe et al. 2015), OpenCourseWare, learning management systems, and electronic teaching materials, as well as dissemination of various education/learning tools making use of IT. Furthermore, we are gathering great expectations as a means to elucidate through visualization of education and learning process, improvement of education through measurement and verification of learning effect, and as a means to support and promote students' autonomous and effective learning.

An exemplary practical example of learning analytics is the open learning initiative (OLI) of Carnegie Mellon University which began in 2003 (Lovett et al. 2008).

In OLI, a teaching/learning model based on cognitive learning theory is applied to the design of online lectures.

The material of each lecture is modularized and can be partially used according to the needs of users, and an intelligent tutoring system for learning support and virtual experiment simulation is also incorporated.

The OLI can navigate a learning route that is optimal for individual learners using the user's learning history.

In addition, by utilizing OLI before in class meetings as part of the students' self-study, we can analyze the results of each learning log and find which concepts were most difficult for the students to understand. This aspect is similar to flipped learning but still has the benefits of face-to-face learning, creating a more effective blended learning environment.

Learning Catalytics, developed by Professor Eric Mazur of Harvard University in 2011, applies a learning method called peer instruction in which "learners mutually teach each other." There, it provides a function to verify the effect of peer instruction by analyzing the response of each student who used a clicker (personal response system: PRS) in the classroom over time. It also has a recommendation function that presents a combination of students to further enhance the possibility of reaching correct answers.

In this way, learning analytics can be used in various teaching/learning modes ranging from online lecture/teaching materials, blended learning and face-to-face lessons, but the scope of application is not limited to lecture/class level.

For example, at a more macroscopic level, it is possible to analyze learners' characteristics, studies, learning, performance data, etc. and use them for evaluation and improvement of educational curriculums (Long and Siemens 2011).

It is useful to examine what form of lesson/teaching material is more effective depending on the characteristics of the learner and to improve education from a structural and system viewpoint.

It is also effective to think about what kind of order the lesson will be in chronologically to improve the learning effect.

We first mention "personalized learning," considering that learning analytics has become more and more important for education and learning in modern society.

Already in various knowledge and information sectors of society, conversion from "Supply Push" to "Demand Pull" has been aimed at individualization of services. Based on educational big data obtained from a larger number of learners, it is desirable to utilize the results in the field of artificial intelligence research and to support diverse and effective education/learning according to individual characteristics and purpose.

In doing so, considering how to bring about "adaptive learning" for individuals through each stage of "descriptive", "diagnostic", "predictive", and "prescriptive" learning analytics is an essential subject.

Learning analytics is used in an exploratory approach of "understanding what education and learning-related data are available for the time being."

However, it is important to clarify for "what purpose to use learning analytics" and to examine, collect, and analyze the data and evidence required qualitatively and quantitatively.

It is also important to think from the viewpoint of "how to present and use analysis results in an easy-to-understand manner."

In modern society which is more complicated and fluidized, the obsolescence of technology and knowledge becomes intense and employment is becoming difficult to stabilize. Therefore, in order to secure the knowledge, skills, and professional foundations necessary for individuals, individuals need a new educational system that allows them to continue learning as necessary for their lifetime.

With the advancement of IT such as the Internet and multimedia technology, digitization of educational tools and contents is rapidly progressing.

Over 1000 paid online degree programs have already been created in the world, and tens of thousands of online lectures/teaching materials are released free of charge. On the other hand, how to effectively utilize these educational resources and environments for learning throughout life will largely depend on the development and diffusion of learning analytics in the future.

1.5 Machine Learning Accelerates Human Learning

At the forefront of learning analytics is "evidence-based education."

The essential technique for that is machine learning.

Records of learners represented by e-portfolios are used for machine learning and can be used to suggest appropriate guidance.

The educational data gathered by learning analytics will be used by the AI tutor as data that are the basis for the learner's proper learning. Specifically, the AI tutor shows the reason for choosing the recommended course of study.

This is to analyze the advantages and disadvantages of learners from the past learning history, to prolong the strengths, to recommend courses to overcome disadvantages, and to explain the reasons.

Machine learning predicts how the learner's ability will improve when that course is selected. In general, machine learning cannot explain how the prediction result was derived, but research on "explainable AI" to make it possible is being performed.

A combination of learning analytics and explainable AI enables evidence-based education. For that purpose, the learner must always be able to confirm the results of analysis and evaluation of various data in the learning process. There are various confirmation methods, and it is possible to visualize data by infographics and to read automatically generated sentences. Mutual evaluation by members belonging to the same group as learners is also effective. By evaluating others, you will be able to view your learning state objectively. The results of the mutual evaluation will also be easy to visualize and can be confirmed.

We are incorporating a mechanism of evidence-based education for training student's discussion skills. In this mechanism, we use a system (discussion mining system) that acquires and analyzes discussion data in detail. In addition, in the manner of gamification, feedback on the progress of learning is easy to understand (gamified discussion).

Regarding discussion mining, as explained in Chap. 2, it is a research method that records, analyzes, and evaluates discussions in the university laboratory in detail. By expressing the relationship between statements by a tree structure, we structure the discussion, give attributes to individual statements, and make meaning of the discussion data. Since discussion in the laboratory is strongly tied to student research content, strengthening discussion skills will lead to promotion of student's research activities.

Regarding the gamified discussion as explained in Chap. 4, we applied gamification to the discussion in which based on the rules, score participants' actions (that is, statements), and compete within the group to train discussion skills. Since the results are fed back for each action, it is easy to understand the state of the participant himself/herself and others, and it is easy to maintain motivation to continue the action. The goal can be decided for each participant, and the list of goals presented as an option is associated with a system (courseware) for learning discussion skills, depending on the current situation for each participant.

Machine learning that predicts the future state based on detailed data also plays an important role in human learning. For example, it is possible to predict the suitability of a learner and recommend a learning course suitable for it.

For that purpose, we must construct a learner model (user model). The recommendation system method can be applied to this. A recommendation system is a system that connects products and services with users.

The user model in the recommendation system should always represent the user's interests and purposes without excess or deficiency. What is necessary for this is the detailed data generation of user behavior.

We collect and analyze various data in educational research activities.

Specifically, it is the discussion data described in Chap. 2. This is a record of all the statements of the student in the meeting (seminar) at the university laboratory.

The created dataset includes audio–visual data exceeding 50,000 times of statement data (total number of participants is over 6000) and 1500 h ($\times$number of cameras 3) at about 700 meetings collected from April 2006.

We also collect data on students' post-meeting research activities using the system called Creative Activity Support System described in Chap. 3. This is a history of actions that the student performed after analyzing the data of the discussion at the meeting to automatically discover candidate tasks to be performed and presented to the students.

The students' activity data include notes organizing the contents of the task, plans for its execution, activity reports, and evaluation results from other students. The plan is designed to clarify the relationship with the overall goal by comparing it with the map that comprehensively covers the research activities (data covering the activities from the start of the research to the presentation of the paper and its goals, achievement criteria, etc.).

By analyzing the data collected by the creative activity support system, the characteristics of each student become clear. It will become clear what kind of task can be accomplished efficiently and what other tasks cannot. For example, although a student is good at system implementation by programming, the student is not good at summarizing research trends by surveying articles. This feature is effective for customizing the learning support system.

An appropriate learning plan for the learner is generated by accumulating behavior data in the learning process and feature analysis based on machine learning. Human learning is promoted by putting this plan into practice. The plan may be changed dynamically, but it can be explained to the learner who will be able to understand the reason at a fundamental level.

Humans understand their characteristics well, set goals, and decide tasks to be performed or subjects to be learned. The support system will guide the learner so as not to deviate from the target while repeating evaluation and adjustment. In this way, human learning is accelerated.

1.6 Deep Learning Approaches

Since Perceptron was invented in 1957, multilayered neural networks with hidden layers (intermediate layers other than input and output) have been actively researched, but due to technical problems such as local optimal solutions it was not able enough to make it learn well and the performance did not rise, so there were questions about its practicality. However, in 2006, a research team of Geoffrey Hinton who is a researcher of neural networks reported that by restricting the hidden layer in multiple layers in the restricted Boltzmann machine (neural network with recursive structure). By devising a method to efficiently learn a multilayered neural network, it came to attract attention again. This research result is seen as a technological breakthrough directly linked to current deep learning. Also, from 2012 onward research has been rapidly active, it is said that a tertiary artificial intelligence boom has arrived.

In deep learning, learning is performed so that the error is minimized by calculating the output error when inputting a large amount of training data to the multilayered neural network by the calculation method called backpropagation. At this time, conventionally, feature amounts manually set by researchers and technicians of respective data such as images and sounds are automatically extracted. For this reason, it is one of the great advantages of deep learning that the feature quantity extraction by hand is almost unnecessary.

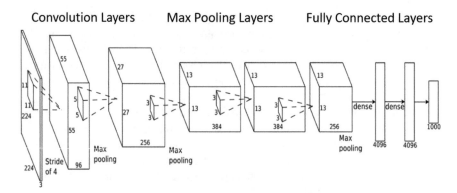

Fig. 1.1 Example of convolutional neural network

Multilayering of neural networks has problems in learning time and calculation cost, but recently it has been difficult to improve the performance of computers and improve the performance of graphics processing unit (GPU), which is superior to parallel processing of operations simpler than a central processing unit (CPU) by general-purpose calculation (GPGPU). By using the GPU, it is said that price per performance and power consumption can be suppressed to 1/100th of a CPU.

The most important techniques in deep learning are convolutional neural networks (CNN) and recurrent neural networks (RNN).

Consider, for example, an all-coupled neural network in which all the nodes of each layer are coupled between layers.

At this time, the number of parameters becomes very large, not only the learning efficiency is deteriorated but also excessive learning (excessive adaptation to the training data and deterioration of accuracy in the test data) is likely to occur. Also, if it is attempted to treat not only one-dimensional vectors but also three-dimensional matrices as one layer (mainly in the case of images), calculation becomes much more troublesome. Considering a filter that can be applied to a specific region of one layer, the value obtained through the filter is taken as the input to the next layer. Accordingly, it is possible to determine the correspondence relationship between the layers only by the parameters set in the filter. This operation is called convolution.

The general shape of CNN used in actual image recognition is as shown in Fig. 1.1. Detailed explanation will be omitted, but when an input image is given, features of the image are extracted by the operation of convolution and pooling, the extracted features are input to the entire combined network, and the class of the image is finally estimated. Here, pooling is rough resampling of the convolution output of the previous layer (for example, taking the maximum value of the value in the $n \times n$ region of the input image, etc.). This makes it possible to absorb the difference in appearance due to some deviation of the image, and it is possible to acquire the invariable feature amount with a slight shift.

The connection pattern between neurons (nodes) in CNN is inspired by the visual cortex of animals. Individual neurons that respond only to stimuli in the limited field

1.6 Deep Learning Approaches

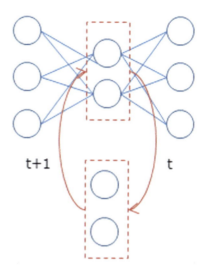

Fig. 1.2 Recurrent neural network

of view are called receptive fields. The receptive fields of different neurons partially overlap to cover the entire field of view.

In this way, CNN is implemented based on various knowledge concerning image processing, and in recent years, CNN has given very high results in object recognition from images.

CNN can process classifications by processing images and sound patterns at a specific time. However, in order to recognize the state from the moving picture and to understand the meaning of the voice, it cannot be said that identification by time is sufficient. Therefore, an RNN that can handle time series information before and after was proposed. An example is shown in Fig. 1.2.

The content of the hidden layer at time t becomes the input at the next time $t + 1$, the hidden layer at $t + 1$ becomes the input at $t + 2,...$ and so on. That is, the state of the previous hidden layer is used also for learning the next hidden layer. Since RNN can be considered to be the same as ordinary neural network when it is expanded in time, backpropagation can be applied to parameter learning as with CNN and the like. The temporal expansion of RNN can be briefly shown in Fig. 1.3.

The error (the difference between the teacher signal and the output) propagates from the last time T toward the first time 0. Therefore, the error at a certain time t is the sum of "difference between teacher signal and output at time t" and "error propagated from $t + 1$." The backpropagation performed in this way is called backpropagation through time (BPTT).

Since BPTT cannot perform learning without data up to the last time T, that is, all time series data, it is necessary to handle long time series data by cutting off the latest time section. There are various problems in this BPTT, and various learning methods to deal with it have been devised. One of them is long short-term memory (LSTM).

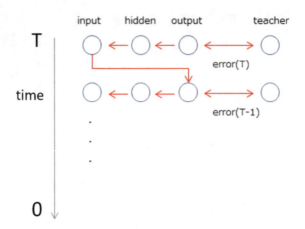

Fig. 1.3 Temporal expansion of RNN

If T is very large, that is, in the case of long time series data, errors propagated due to calculation problems become very small or conversely very large. Although it can cope with the maximum value restriction for the larger value, it is difficult to deal with it becoming too small. Therefore, LSTM incorporates a method of propagating error so as not to attenuate greatly.

Deep learning such as CNN and RNN is very effective mainly in pattern recognition such as speech recognition and image processing. As described in Chap. 3, it is used for converting speech into texts and recognizing facial images of participants. Currently, we do not use deep learning techniques to analyze various data in human creative activities. However, when the data are on larger scale and it becomes difficult to manually perform feature extraction for modeling, we must rely on deep learning methods. However, since machines cannot be taught correctly for problems that humans cannot show correct answers, it is necessary for humans to understand the essential mechanism for promoting human learning well.

References

S. Bull, J. Kay, Student models that invite the learner in: the SMILI open learner modelling framework. Int. J. Artif. Intell. Educ. **17**(2), 89–120 (2007)

J. Carbonell, AI in CAI: an artificial-intelligence approach to computer-assisted instruction. IEEE Trans. Man-Mach. Syst. **11**(4), 190–202 (1970)

W. J. Clancey, *Knowledge-Based Tutoring: The GUIDON Program* (The MIT Press, 1987)

A. Collins, J. S. Brown, S. E. Newman, in *Cognitive Apprenticeship: Teaching the Craft of Reading, Writing, and Mathematics*. Technical Report 403 (BBN Laboratories, Cambridge, MA, Centre for the Study of Reading, University of Illinois, 1987)

A. Kashihara, Modeling learning in engineering. J. Jpn. Soc. Artif. Intell. **30**(4), 473–476 (2015). (in Japanese)

P. Long, G. Siemens, Penetrating the fog: Analytics in learning and education. EDUCAUSE Rev. **46**(5), 31–40 (2011)

References

M. Lovett, O. Meyer, C. Thille, The open learning initiative: measuring the effectiveness of the OLI statistics course in accelerating student learning. J. Interact. Media Educ. **1**, 2008 (2008)

K. Nakabayashi, Technical standardization trend of educational support systems. J. Jpn. Soc. Artif. Intell. **17**(4), 465–470 (2002). (in Japanese)

D.A. Norman, Learner-centered education. Commun. ACM **39**(4), 24–27 (1996)

K. Nakabayashi, e-Testing and standardization, e-Testing (Baifukan Co., Ltd, 2008), pp. 74–94. (in Japanese)

K. Nakabayashi, Standardization of e-learning technology and design of learning activities. J. Jpn. Soc. Artif. Intell. **25**(2), 250–258 (2010). (in Japanese)

R. Nkambou et al., *Advances in Intelligent Tutoring Systems (Studies in Computational Intelligence)*, vol. 308 (Springer, Berlin, 2010)

X. Ochoa, D. Suthers, K. Verbert, E. Duval, Analysis and reflections on the third learning analytics and knowledge conference (LAK 2013). J. Learn. Analytics. **1**(2), 5–22 (2014)

F. Watanabe, Y. Mori, C. Kogo, Analyzing learners' subjective evaluation of peer assessment in japan massive open online courses, Waseda. J. Human Sci. **28**(2), 237–245 (2015)

E. Wenger, *Artificial Intelligence and Tutoring Systems: Computational and Cognitive Approaches to the Communication of Knowledge* (Morgan Kaufmann Publishers Inc., 1987)

B. Woolf, *Building intelligent interactive tutors* (Morgan Kaufmann, Burlington, MA, 2009)

Chapter 2
Discussion Data Analytics

Abstract Evidence-based research, such as research on big data applications, has been receiving much attention and has led to the proposal of techniques for improving the quality of life by storing and analyzing data on daily activities in large quantities. These types of techniques have been applied in the education sector, but a crucial problem remains to be overcome: it is generally difficult to record intellectual activities and accumulate and analyze such data on a large scale. Since this kind of data is not possible to compress in a manner, such as taking the average, it is necessary to maintain the original data as the instances of cases. Such human intellectual-activity data should be treated as big data in the near future. We have been developing a discussion mining system that records face-to-face meetings in detail, analyzes their content, and conducts knowledge discovery. Looking back on past discussion content by browsing documents, such as minutes, is an effective means for conducting future activities. In meetings at which some research topics are regularly discussed, such as seminars in laboratories, the presenters are required to discuss future issues by checking urgent matters from the discussion records. We call statements including advice or requests proposed at previous meetings "task statements" and propose a method for automatically extracting them. With this method, based on certain semantic attributes and linguistic characteristics of statements, a statistical machine learning model is created using logistic regression analysis. A statement is judged whether it is a task statement according to its probability. We also developed a method that maintains the extraction accuracy by using the discussion mining system and its extension on the basis of task statement extraction over a long period. Specifically, we constructed an initial discriminant model of task statements and then applied active learning to new meeting minutes to improve the extraction accuracy. Active learning also has the advantage of reducing labeling costs in supervised machine learning. We explain the improvement in extraction accuracy and reduction in labeling costs with our method and confirm its effectiveness through simulations we conducted.

Keywords Discussion mining · Discussion structure · Summarization · Task discovery · Statistical machine learning · Logistic regression analysis · Active learning

© Springer Nature Singapore Pte Ltd. 2019
K. Nagao, *Artificial Intelligence Accelerates Human Learning*,
https://doi.org/10.1007/978-981-13-6175-3_2

2.1 Structure of Discussion

We can think of discussion as a form of communication with some form of rules attached. I believe this because in comparison to daily conversation, there are many more things we must think about and decide ahead of time in regards to what we plan on discussing. For example, should I expand on the current topic or conclude my point and prepare to move to another topic. Therefore, we decided to divide the remarks in the discussion into two types such as "start-up" and "follow-up". Start-up remarks are those that speak about new topics, and follow-up remarks are to continue speaking on the current topic. This classification is done manually at first, but it can be done automatically by applying a machine learning technique.

As another constraint, we decided to regard the discussion as a collection of tree structures. A tree structure is a data structure that hierarchically derives from a single point called a root to multiple points (nodes). One tree structure expresses the structure of discussion on one topic. At this time, each point of the tree structure, that is, a node is a statement or a remark, and a link connecting a node and another node is a relation between statements.

By using a tree structure, various calculation algorithms can be applied. For example, by sorting nodes based on the distance from the route or by comparing the number of nodes that have the same node as a parent (sibling nodes having a direct relationship to the route), it is possible to select a certain topic to investigate more deeply.

We also associate multiple discussions by generating a graph structure with links between multiple tree structures, using the materials (presentation slides like PowerPoint) used in those discussions. The difference between the tree structure and the graph structure is that there is no route in the graph in general, and there is not necessarily a parent node for each node (to be precise, the tree structure is a graph structure with some restrictions).

By structuring the discussion in this way, it can be used as the data to be subjected to scientific analysis. There are various things that you miss if you catch the discussion as a collection of just statements. For example, what relationships exist between which statements? If a person asks a question and another person answers it, there is a clear relationship between the question and the answer. Another person sometimes gives a supplementary explanation to someone's statement. Between these statements, there is a relationship that the latter supplements the former (adds information). Structuring the discussion is to put a relationship between the statements in this way.

It can be said that the basic ability of discussion and thus communication is the ability to have such statements related to other statements. If this ability of someone is low, that person will confuse the surroundings and will lead to spend time wasting just to say things that are not much related to others.

However, it is inappropriate to say only opinions to other people's opinions, such as "I think so". Except for deciding by majority vote, it is often insufficient to express only the intention to agree. Again, you should say something like adding information. In such cases where you only agree, it is better not to speak, but just to add attributes to the remarks to be agreed. The structuring of the discussion includes not only relating

2.1 Structure of Discussion

other statements to the statement but also attaching attributes to some remarks (e.g., agree or disagree).

Attributes for the statements include evaluation of the statements. In other words, we add points (scores) to the statements of others. This is different from showing the attitude of agreeing or opposing the statements, judging whether the quality of the statement is high and attaching evaluation attributes.

By saying something related to a certain statement, or by listening to and evaluating certain statements and so on, the structure of the discussion becomes articulate, and the speakers' ability of discussion will rise. Therefore, the structure of discussion is effective not only for humans but also for machines, and machines will support humans. There is a technology called data analytics which is used to make machines utilize structured data. It is also a technique for humans to better understand the characteristics of these data.

There is also a technology called natural language processing for analyzing human language. We also use natural language processing and data analytics to analyze and evaluate discussions.

Now, let's take a closer look at the mechanism that we invented.

2.2 Discussion Mining System

Seminar-style meetings that are regularly held at university laboratories are places where exchanges of opinions on research content occur. Many comments on future work are included in their meeting records. However, as discussions at meetings are generally not recorded in detail, it is difficult to use these for discovering useful knowledge. Our laboratory, on the other hand, has been developing and operating a system that systematically records the content of face-to-face meetings with metadata and achieves support, such as that from reviews of discussion content (Nagao et al. 2005). Although it is essential to review tasks to set new goals in research activities, their existence may be concealed in many other statements in the minutes.

Therefore, we have proposed a system to support task execution in student's research activities by combining a mechanism called discussion mining (hereinafter referred to as DM) that realizes data mining on the contents of the discussion and a machine learning method called active learning, and we have developed a mechanism to operate it for a long time.

In our laboratory at Nagoya University, we have used this DM system to record detailed meetings in the laboratory for over ten years. This system enables all participants to cooperate to create and use structured minutes. This system is not fully automated, i.e., the secretary manually writes the contents of the speech, and each speaker tags his/her speech. Therefore, we can generate data with high accuracy.

As mentioned at the beginning of this chapter, there are two kinds of tags given to remarks, one tag is a newly introduced topic to introduce new topics, and the other is a tag of continuous talk continuing the topic already discussed. In the case of continuing utterances, it is necessary to clarify from which remarks they continue.

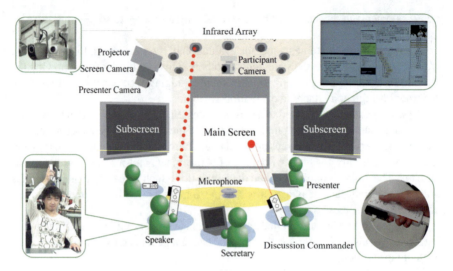

Fig. 2.1 Physical configuration of discussion mining system

We see that discussion takes on a tree structure form as the speakers always attach attributes to their speech and the related speech of others. This tree structure is visualized in real time during the meeting and it is used to overlook the discussion.

The meeting style supported by the DM system is that the presenter explains a topic while displaying slides, and question-and-answer with the meeting participants is either conducted during or at the end of the presentation.

Specifically, using multiple cameras and microphones installed in a discussion room, as shown in Fig. 2.1, and a presenter/secretary tool we created, we record the discussion content. In the center of the discussion room, there is also a main screen that displays the presentation materials and demonstration videos, and on both sides, there are subscreens for displaying information on and images of the participants who are currently speaking.

The DM system records slide presentations and question-and-answer sessions including participants while segmenting them in time. As a result, content (discussion content), as shown in Fig. 2.2, is recorded and generated.

Every participant inputs metadata about his/her speech using a dedicated device that is called a discussion commander, as shown in the lower right of Fig. 2.1. Participants who specifically ask questions or make comments on new topics assign start-up tags to their statements. Also, if they want to speak in more detail on topics related to the immediately preceding statement, they provide a follow-up tag. Furthermore, the system records the pointer location on slides, designates the location/time for the slide and information, and has a button for agreeing or disagreeing in relation to statements made during the presentation and question-and-answer session. Marking information on important statements is also recorded.

2.2 Discussion Mining System

Fig. 2.2 Structured discussion content

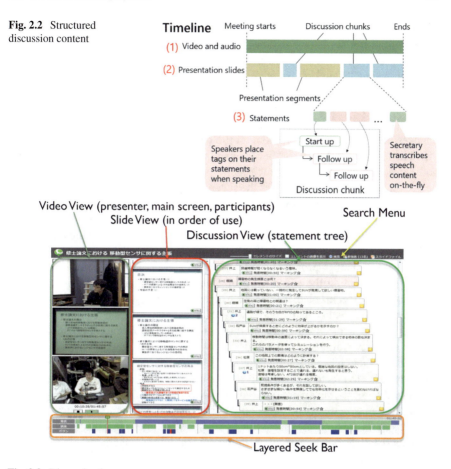

Fig. 2.3 Discussion browser

We also developed a system for searching and viewing recorded data. In this discussion-content-browsing system, a user can search the contents of the agenda from the date and participant information, view past discussions like the ongoing debate, and effectively visualize the state of the discussion, as shown in Fig. 2.3.

2.3 Structuring Discussion

Methods of acquiring metadata about meetings include a method using an automatic recognition technique, such as a meeting browser (Schultz et al. 2001) and a method of human input using devices and tools such as a conversation quantizer (Nishida 2007). The amount of human effort when acquiring metadata is very small with the former method, but it is currently difficult to automatically record all necessary information on a computer. Although it is possible to search for the keywords of statements, they do not contain sufficient information on the content to be understood when browsing. Therefore, we adopted a method in this research in which humans and machines cooperatively input these metadata.

In the DM system, the presenter uploads a presentation slide by using a special tool and the system automatically transmits the slide information when the presenter changes the slide page. In addition, all the participants use the portable discussion commander. The start time for the statement and the statement type are recorded, in addition to the speaker ID and the seating position of the speaker, by placing this device on the top. The end time for the statement is input by pressing a button on the device. The system segments video/audio information on each statement by acquiring the start/end times of the statement. In addition, it is possible for a user to express an attitude (agree or disagree) with respect to the statement by pressing a button on the device or to mark a statement that has an important meaning to himself/herself.

The DM system is provided with a statement reservation function as a mechanism for controlling the order of statements by participants to prevent crosstalk. When someone signals his/her intention to speak by holding the discussion commander above them, his/her name with the type of statement is added to the statement reservation list, and his/her speaking turn automatically shifts to him/her when the immediately preceding statement ends.

This statement reservation function is not only used to control the order of statements but also to create a discussion structure that reflects human intentions. If the system only uses the statement type without using the statement reservation function, the discussion structure that is created will be a list structure starting from the start-up statement. However, the discussion structure when multiple participants express opinions from various perspectives on the content of one statement is a tree structure rather than a list structure. Therefore, when a follow-up reservation is added while someone is speaking, the DM system generates link information between the follow-up statement and the ongoing statement. In other words, when multiple reservations are added during speaking, a tree structure in which a plurality of follow-up statements is repeated is automatically created for one statement.

The root of the tree structure in this case is a start-up statement, where all the others are follow-up statements. If several follow-up statements are added at the same time for one statement, the branches of the tree increase. Follow-up statements are attached to the preceding statement, which deepens the tree structure.

In addition, the secretary records the contents of the statements using a special tool. This tool is linked with the discussion commander of each speaker, and when the speaker starts speaking, the "speaker and speech type" node is automatically generated in the secretary tool. By selecting this node, the secretary can efficiently record the contents of the speech during the discussion.

2.4 Summarization of Discussion

We consider important utterances to have a large impact on the thoughts and opinions of participants, to encourage active discussion and to skillfully summarize ideas from the discussion up until that point. What can be considered as factors that affect the importance of statements are the number of branches of that statement (the number of follow-up statements following the statement), whether it is on a longer thread, the social position of the speaker, and the cumulative number of follow-up statements.

Although we wanted to determine what statements were important, start-up statements were not always important, and statements by specific speakers were not always consistently important.

Therefore, we propose an algorithm to discover important statements by spreading activation (Nagao and Hasida 1998). This is based on a network where nodes are statements and links are introduced based on the structure of the discussion segment (start-up/follow-up tags) and links based on the pointer referents (links of statements having the same referred object are linked).

Spreading activation is a model in which the activity value of a node diffuses to a neighboring node through a link in a network represented by nodes and links, as outlined in Fig. 2.4. In other words, the activation values of nodes closer in distance to a node with a higher activation value are also higher, and conversely, a node whose distance is farther is also lower in activation value. By applying this method, it is possible to calculate the relative activation value of each node for all the nodes in the network.

Specific methods of calculation are as follows. First, we created a matrix A that represents the presence or absence of a link between each node by assuming that the number of nodes existing in the network is n.

Here, the value of (i,j) is one if there is a link between nodes i and j, and zero if it does not exist. We now create matrix W that is related to the weight of spreading activation by using matrix A. Matrix W is an $n \times n$ matrix obtained by dividing the value of each row of A by the number of nonzero elements of that row (this is done to

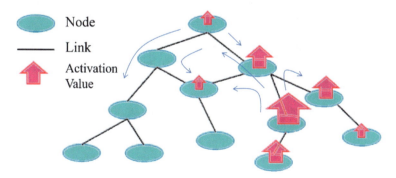

Fig. 2.4 Spreading activation

converge spreading activation). Matrix C that represents a constant activation value spreading from the respective nodes to the whole network is next introduced (red arrow in Fig. 2.4).

Here, the value of c_i is the external input value of the constant given to node i.

Then, by solving the recurrence formula, $X_0 = CX_{t+1} = W \cdot X_t + C$, the activation value in the network at the time of step t (=0, 1, 2) is obtained. Here, $x_{t,i}$ represents the activation value of node i at step t. Finally, the activation value of each node in the network is calculated as $t \to \infty$. Also, when $t \to \infty$, the above recurrence formula can be expressed as $X = W \cdot X + C$ so that $X = (E - W)^{-1} \cdot C$, where E represents an identity matrix).

We implemented a mechanism on the browser that selects important statements according to the purpose of browsing to efficiently browse the recorded discussion. We specifically constructed a network where nodes from each statement from the link information were obtained from the discussion structure and pointing acts, calculated the activation value given to each node from the agreement/opposite button and marking, and used the spreading activation algorithm. We then ranked importance ranking and filtered the statements.

For example, if the follow-up statement of statement i is j, there is a link between statement i and statement j. By repeating this, a network of statements is formed for each discussion segment starting with the start-up statement. Further, the links of all the start-up statements in each discussion segment are linked with a certain virtual node. All the discussion segments recorded from this are represented by one network structure. We can then execute the spreading activation algorithm on this network structure and find the relative activation value for each statement. Values that take into consideration various metadata acquired at a meeting are used as external input values to be given to each node. Specific metadata include statement types, button information for agree/disagree, marking information, speaker names, and keywords included in the object. Since the start-up statement is a premise of follow-up statements in the same discussion segment, it is set so that the activation value is higher than that of the follow-up statements.

The users of the discussion browser (Fig. 2.3) use the user interface to find how much importance is given to each item of metadata (Fig. 2.5). They can interactively use this interface and discover a set of statements that is suitable for their own reading purposes. For example, if a user increases the importance of the start-up statements, he/she can efficiently discover the statement set that has the same topic as the start-up statements, and the statement set that gained numerous agreements when the importance of the agree button information was increased by the user.

By using this system, the user can view the summary of the discussion and the video of each statement. We believe that the discussion that we usually do is very important, so we will try to make effective use of minor content as much as possible. However, as content increases, it will become more difficult to look at each discussion in detail. Even though it is possible to automatically extract important parts by summarization, it is still troublesome to look across multiple meeting contents. Therefore, it is necessary to have a mechanism that automatically discovers useful information from the discussion.

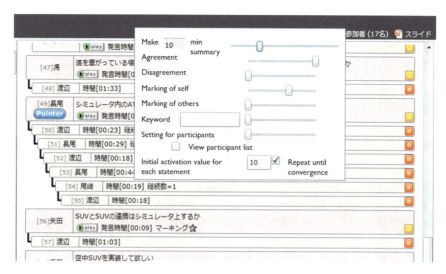

Fig. 2.5 Parameter setting window in discussion browser

2.5 Task Discovery from Discussion

Remembering past discussion content helps us to seamlessly carry out future activities. For example, presenters in laboratory seminars can remember suggestions and requests about their research activities from discussion content that has been recorded in detail. The meeting content contains useful information for the presenters, but it is onerous to read the information. As necessary information is concealed in a large quantity of statements, it is not easy to find. This is problematic if past discussions are not being reviewed, even for other speakers, and not only presenters. Therefore, it is necessary to extract information that concerns unsolved issues from previous discussions. We have called statements that include future tasks "task statements".

We developed a method of statistically determining whether statements were about future tasks, i.e., task statements (Nagao et al. 2015). Some attributes, including linguistic characteristics, structures of discussions, and speaker information, were used to create a probabilistic model.

A task statement can include any of three types of content:

1. Proposals, suggestions, or requests provided during the meeting: the presenter has determined that they should be considered.
2. Problems to be solved: the presenter has determined problems that should be solved.
3. Tasks not yet carried out before the meeting: the presenter has occasionally already noted such tasks.

Candidates of task statements are fragments of a discussion chunk, which was explained earlier. A typical chunk is created from one or more of the questions and

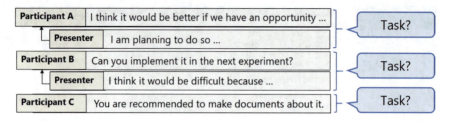

Fig. 2.6 Candidates for task statements

comments of the meeting participants and the presenter's responses to them. A coherent piece of discussion content related to tasks consists of questions/comments and their responses. Thus, "participants' questions/comments + presenter's responses" are a primary candidate and a target of retrieval. "Participants' questions/comments and no response" is a secondary candidate. Figure 2.6 has an example of candidates for task statements.

Below, as an example of task statements manually discovered from our discussion, we list participants' statements and their response to them:

(1a) I think that there is a mechanism of dynamic prediction, such as using the phrase in the PowerPoint material.
(1b) I really want to use the current slide.
The statement 1a advises the presenter and the speaker in the statement 1b expresses his willingness to do "I really want to …", which applies to Condition 1 described earlier.
(2a) I am glad if there is a function that can move immediately by pressing a specific number.
(2b) If you enter a number as it is, it conflicts with other inputs and when pressed together with Shift, etc., it seems to conflict with other shortcuts.
In statement 2a, participants inform the presenter of the request, while in statement 2b the speaker mentioned the problem in realizing the request and it has not been decided that it should be implemented at this point. Condition 2 is applicable because it includes content to consider whether to implement or how to implement it.
(3a) If you are targeting students of Nagao Lab, you think that you are not accustomed to typing in predictive transformations, so make opportunities to get used to that predictive transformation, and think that if you use the secretary tool with familiarity it will be used so much. How about that?
(3b) I plan to have a chance to get used to it.
The statement 3a states that the participant suggests to the presenter in remark 3a and the speaker in statement 3b states that the proposal is scheduled to execute, and Condition 3 can be applied to it.

Discovery of task statements is the result of the so-called supervised learning where a computer learns by giving information for discrimination (called teacher

2.5 Task Discovery from Discussion

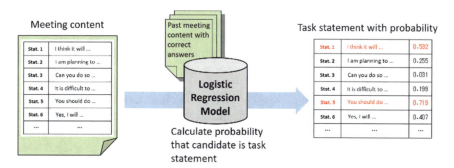

Fig. 2.7 Overall process of task statement extraction

signals) by humans, and a set of sample data for discrimination of past discussions (called a training dataset).

Using the past data of the discussion as training data, we will discover the task statement by the following algorithm. More precisely, we assign the probability of judging that it is a target utterance of a set of candidate utterances.

1. Probabilistic model based on logistic regression analysis is created using correct answer data (teacher signal) created by hand. Model parameter estimation is performed in the manner described in the following section.
2. Calculate the probability that the set of statements in the latest discussion is a task statement by using the created probabilistic model.
3. Extract statements whose probability value exceeds a threshold (for example, 0.5) as task statements.

This procedure is shown in Fig. 2.7.

In generalized linear regression methods including logistic regression analysis, there are concepts such as regularization for suppressing over learning (also called overfitting, over adaptation to training data and failing to adapt well to other data). In addition, we adopted the logistic regression analysis for this problem because it has high scalability and can be easily combined with other active-learning methods I will mention later.

Some of the past meeting content was manually analyzed to find characteristics that could be used as clues to extracting task statements. The survey data included 11 types of meeting content and 598 groups of statements (candidates). Each presenter at the meetings manually selected task statements from each type of content.

As a result of manually extracting task statements in the survey data, 246 task statements were found, which corresponded to 41.1% of all candidates. We analyzed the characteristics of the task statements by comparing such percentages. For example, statements made by teachers had a higher overall percentage of task statements. Therefore, speaker attributes were helpful in calculating the probabilities of task statements.

As was explained earlier, presenters used their discussion commanders to mark statements that they wanted to check later during the meetings. We investigated the effectiveness of marking for discriminating task statements by calculating the percentage of marked task statements in all task statements. The percentage of task statements that were marked were 73.4%, which was higher than that of the task statements for all candidates.

We found distributions for the respective characters of a presenter's and participants' statements to examine whether there were characteristic tendencies in the number of letters (characters) in task statements. We divided the number of characters into five groups and calculated the percentages of task statements in each group. The percentage of task statements in the participants' statements increased when the number of characters increased. This was because the number of characters in their statements increased when the participants were making concrete requests and giving advice. In comparison, the number of characters with higher percentages of task statements was 20 or less for the presenter's statements. The more characters there were, the smaller the percentage of task statements. We believe that if a presenter accepted the requests or advice from participants who were present, his or her responses would have tended to be brief.

We also investigated the types of sentences included in the task statements. The percentage of task statements in the participants' statements was higher when sentences were in present tense and in declarative form (56.1%). This was due to the fact that a large amount of advice or requests were in the pattern of "should be …" or "I want to …". The percentage of task statements for the presenter's statements in past tense and in declarative form was low (29.2%). This was because presenters did not tend to use sentences in past tense when they talked about future tasks. In addition, the percentage of task statements made by presenters in past tense and in interrogative form was 0%.

Morphemes and collocations of morphemes in statements were also important features. We generated a morpheme bigram of nouns, verbs, adjectives, and auxiliary verbs in the survey data by calculating the number of occurrences of the morphemes. We detailed this topic later.

We then determined a feature of morphemes and their bigrams of statements if their occurrences had exceeded certain thresholds. The selected nouns specifically had a percentage of occurrences that was greater than or equal to 0.5% of all nouns, and the selected verbs also had a percentage that was greater than or equal to 0.5% for all verbs. Morpheme bigrams were selected if their percentages were greater than 0.05% for all morpheme bigrams. These selected morphemes and bigrams were used as features for discriminating task statements.

Three main features were selected based on these results obtained from the survey to create a prediction model.

1. Attributes of presenter.
2. Features of participant's statements:

    (a) Start time and duration of statements,
    (b) Speaker type (teacher or student),
    (c) Statement types (start-up or follow-up),
    (d) Marking (zero or one),

2.5 Task Discovery from Discussion

(e) Length (no. of characters),
(f) Sentence types,
(g) Morphemes and morpheme bigrams, and
(h) Responses by presenter (zero or one).

3. Features of presenter's response:

(a) Marking (zero or one),
(b) Length (no. of characters),
(c) Sentence types, and
(d) Morphemes and morpheme bigrams.

We used answers (zero or one) to five questions for the values of sentence type features:

1. Does the statement include a sentence in past tense and in declarative form?
2. Does the statement include a sentence in present tense and in declarative form?
3. Does the statement include a sentence in past tense and in interrogative form?
4. Does the statement include a sentence in present tense and in interrogative form?
5. Does the statement include a sentence of another type?

Regarding the start time of the statements, we divided the time of the whole meeting into five sections every 20% and used the value in which section the participant's speech start time occurred.

Regarding the number of words in the statements, the distribution of the number of characters of each participant's speech and the speaker's statements were found and divided into five sections every 20%, and the value for the number of characters in the section the speech is in is used.

In addition, we used the answers of the following items as values for the types of sentences included in the statement.

Whether the statement only appears in other sentences.

Whether it is a statement that ends with a noun or noun phrase or not.

Whether or not it is a statement that is just composed of a noun or noun phrase
Whether it is a statement that doesn't end on a noun or noun phrase, or does end on a noun phrase, or is just a statement.

Whether or not it is a sentence that does not end on a noun phrase.

One feature is represented as a set of variables with values of 0 or 1 when vectorizing. This variable is called a dummy variable. In other words, when a feature has m categories (cases), it is represented by m variables. This ($i = 1, \ldots, m$) is a dummy variable. If it corresponds to the ith category, it is represented by ($i \neq j$). Since m dummy variables have redundancy, $m - 1$ dummy variables are usually used. For example, when we characterize gender, there are only two cases of male or female, so we have 1 dummy variable, 0 for male and 1 for female. An arbitrary real number can be divided into several areas and can be represented by the number of the area minus one dummy variable.

The creation of a probabilistic model for judging such task statements is based on supervised learning where human beings learn by giving discrimination classes teaching signals to machines. And its accuracy depends on the training dataset used for learning. Of course, if there is a clear difference compared with general statements, the task statement may be able to create a discrimination rule without doing machine learning. However, if the content of the statements is diverse, it is difficult to apply the statement to a specific pattern.

Tenfold cross-validation was applied to the extraction results to test and confirm the effectiveness of the proposed method. The data used for verification included 42 types of meeting content and 1637 groups of statements (candidates). Each presenter created correct data for task statements for each type of meeting content as well as the survey data provided earlier. The data used for verification were completely different from the survey data.

We confirmed the effectiveness of the proposed method in terms of high precision (index for extraction accuracy), recall (index for extraction leakage), and the F-measure (harmonic mean of precision and recall).

The extracted results from the task statements with the proposed method were a precision of 75.8%, a recall of 64.2%, and an F-measure of 69.5%. In comparison, the results for the three alternative methods of extraction were as follows. The selection of statements that were marked by the presenter had the highest precision (68.9%), and that of the statements from teachers or statements that were marked by the presenter had the highest recall (44.1%) and F-measure (48.7%). The approach we proposed obtained the highest values compared with these other extraction methods.

As was previously explained, the proposed method could calculate the probabilities of candidates for a task statement by using a generated probabilistic model. A candidate whose probability value exceeded a certain threshold was extracted as a task statement. We first set the threshold value to 0.5. It was not guaranteed that this value would be optimal. Therefore, we reevaluated the outputs of the system by lowering the threshold by 0.1 from 0.5. As we found that the F-measure at a threshold of 0.4 was highest (71.4%), task statements should be extracted in the future by setting the threshold to 0.4.

Since data analytics is aimed at deriving generality that cannot be simply found from actually obtained data, human beings teach answers to machines on a steady basis as training data, and machines learn gradually by more accurate parameter estimation. We must wait for it to become smarter.

Therefore, to improve discrimination accuracy, it is necessary to increase the amount of training data. However, it is very labor intensive to give a teacher signal to all statements of the enormous discussion recorded by the discussion mining system. Also, it would be desirable for the presenters themselves to be responsible for teacher signaling tasks that should best understand the content of the discussion. In this way, giving a teacher signal is a task requiring specific human knowledge, and it takes time. I will explain how to solve this problem later.

2.6 Active Learning for Improving Supervised Learning

Generating a discrimination model of task statements is based on supervised learning, where a machine learns from humans who provide a discrimination class of teacher signals to the machine, and the extraction results depend on the training dataset on which machine learning was executed. Therefore, it is necessary to increase the amount of training data to improve extraction accuracy especially when the amount of usable data is relatively small, as it was in this research; however, the assignment of teacher signals to the statements of all minutes recorded by DM is very costly. In addition, it is preferable for the presenter who best understands the content of the presentation to be in charge to minimize the number of misjudgments from teacher signals, i.e., it is also a task that requires specific human knowledge. Teacher signals of task statements cannot easily be generated.

Another issue in supervised learning is feature changes in objects to be extracted over time. There is no problem if the characteristics of the extraction target are completely invariant, but as new students enter the lab and progress with research activities transforms each year, the characteristics of task statements to be extracted change over time.

The problem with feature changes over time has often been discussed. A spam mail filter is a good example. The techniques of spam mail are becoming increasingly more sophisticated, and unless the discrimination model is updated, it cannot adapt to the characteristics of new spam mail, which decreases the accuracy of discrimination (Georgala et al. 2014).

One concern with text minutes in DM is that there are differences in wording due to there being different secretaries who manually input text. Since strict rules are not defined for secretaries, the wording depends on the discretion of each person in charge, and the degree of sentence summarization also differs. Since the language information included in statements mostly affects extraction accuracy in the task statement extraction model, it is necessary to focus on feature changes due to differences in wording.

The discriminant model should always be updated when the amount of data to be analyzed is increased. However, it is very difficult to label all data when there is an increasing amount of data, and labeling incurs large costs. We used active learning (Settles 2010, Shimodaira 2000, Sugiyama and Kawanabe 2012, Liu et al. 2015) to solve this problem, which was used to attempt to improve extraction accuracy for the limited dataset we obtained.

Active learning can be implemented as an algorithm in which a sample that makes the greatest contribution to updating a discrimination model from a large number of samples without teacher signals is selected by a machine, and a teacher signal is given to it to minimize human effort. This process efficiently updates models. All statements in task statement extraction in the minutes of a seminar correspond to samples without teacher signals; some groups of statements are selected using an active-learning method, and the selected statement group is assigned teacher signals by students who are presenters at the corresponding seminar. The research-activity-support system that will be described later encourages students to reflect on the seminar after their presentations and obtain feedback on the automatically selected task statements.

Applying this active-learning method to task extraction in this way both simultaneously solves problems of cost in teacher-signal assignment and feature changes in task statements over time.

Active learning is a technique frequently used in several fields such as natural language processing and biostatistics since expert knowledge concerning the assignment of teacher signals is required and the costs of data collection and teaching-signal extraction are very high. The technique is also suitable for improving the accuracy of task statement extraction in which the cost of teacher-signal assignment is also very high.

A five-step procedure is repeated to apply the active-learning technique to DM:

1. Create a task statement extraction model using a set of statements with teacher signals to determine whether or not it is a task statement.
2. After a presentation has been made at a seminar, carry out task statement extraction by using the recently created minutes and calculate the probability value that each statement is a task statement.
3. Select the target set of statements to which the teacher signal is to be assigned by active learning.
4. Display the results for the selected statement set and ask the system to provide teacher-signal feedback to the user.
5. Add the statement set with teacher signals that have been obtained to the training data of machine learning.

By updating the task statement extraction model by repeating these five procedures, it is not only possible to improve the accuracy of extraction by increasing the amount of training data but also to constantly adapt to feature changes of task statements over time.

Information density is a representative algorithm of the active-learning technique that takes into account the density of the feature space as a sampling reference (Settles and Craven 2008). The strategy with this algorithm is to consider that there is a large amount of information with higher density data; therefore, the benefits of providing a teacher signal are thought to be great.

Uncertainty sampling in giving a teacher signal to the most ambiguous classification is also a typical method of active learning (Lewis and Gale 1994). For example, considering the classification problem as shown in Fig. 2.8, the data to which the teacher signal should be given are different between information density and uncertainty sampling. In other words, data close to the classification boundary are selected for uncertainty sampling, but with information density, data in which there are many data around (data with high data density) are selected.

Since information on teacher signals is not used for density calculation, it is possible to use a large number of samples without teacher signals for learning. As the DM project has been continuing for about 10 years, and the accumulated content of discussion is huge, it is very convenient to apply this method.

2.6 Active Learning for Improving Supervised Learning

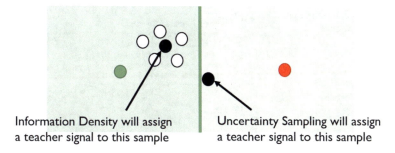

Fig. 2.8 Two typical methods of active learning

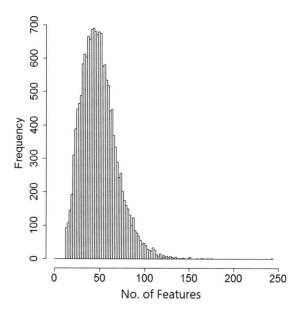

Fig. 2.9 Feature histogram

Although the practicality of information density is high when there is a large amount of noise data, it may adversely affect the weighting of model parameters. Morpheme information on statements is used in the task statement extraction model, and extremely long or short statements can be noise. For example, short statements such as "I will consider it" and "I do understand" that frequently appear in responses by presenters have fewer morphemes that are selected as modeling features, and very long statements that have many morphemes also tend to be noisy samples.

Our proposed method used a weighting algorithm in this research to cope with such problems that was based on a histogram of feature information. Since all the features used in the task statement extraction model are binary variables, we can define a histogram in which the number of features that have the value one in a sample is classified as a class. We call it a feature histogram. Figure 2.9 plots the feature histogram created for the past 492 meeting minutes.

Fig. 2.10 Comparison of accuracy transition

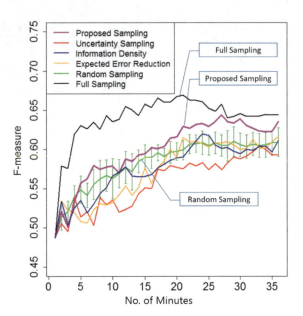

Our weighting algorithm was based on the ratio of the frequency of the number of samples without teacher signals as a weight as

$$\arg\max_{x \in U} \phi_x \times \left(\frac{1}{U} \sum_{i=1}^{U} \text{sim}(x, x_i) \times \frac{\text{freq}(\text{bin}(x))}{|U|} \right) \quad (2.1)$$

Here, ϕ_x is the score of sampling, U is the set of samples without teacher signals, and $\text{sim}(x, x_i)$ is the cosine similarity between x and x_i. The freq() is the frequency of a suggested bin, and bin(x) is the bin to which x belongs. $\frac{\text{freq}(\text{bin}(x))}{|U|}$ is a new weighting, and the other part of Formula (2.1) is the same as the calculation for information density. This weighting makes it possible to give a lightweight to statements of extreme length that can have a negative effect on information density, and more effective sampling to improve extraction accuracy can be expected.

Since teacher-signal feedback improved the accuracy of task extraction, we performed a simulation of what kind of accuracy transition could be observed depending on different data-sampling methods for active learning for a situation in which ten statements (for full sampling of all statements) were added as training data when one minute of the meeting was created. The transitions in the value of the F-measure (or F1 score that is the harmonic mean of precision and recall) were compared with six methods in a sevenfold cross-validation (Fig. 2.10) for the data of the minutes recorded by the DM system (data group: 1637).

- Proposed sampling: formula (1) (proposed method).
- Uncertainty sampling: conventional method (Lewis and Gale 1994).

2.6 Active Learning for Improving Supervised Learning

37

- Information density: conventional method (Settles and Craven 2008).
- Expected error reduction: conventional method (Roy and Mccallum 2001).
- Random sampling: statements are randomly selected for sampling.
- Full sampling: all statements are selected to assign teacher signals.

Full sampling indicates the limits of improving extraction accuracy as a reference value, where the amount of training data is about four times that of the other methods. The proposed method can maintain high extraction accuracy compared with the method excluding full sampling by reducing noise on the feature space, which has not been considered in conventional methods.

2.7 Natural Language Processing for Deep Understanding of Discussion

In the analysis of discussion, morpheme unigram and morpheme bigram are each used as features. Specifically, for morpheme unigrams, the number of occurrences of each of nouns, verbs, adjectives, and auxiliary verbs was calculated from past minutes, and those that exceeded a certain value were taken as features. Specifically, I used as features, nouns and verbs that occurred with a ratio of more than 0.5% out of all nouns and verbs, and adjectives and auxiliary verbs that occurred with a ratio of more than 1.0% out of all adjectives and auxiliary verbs. For morphological bigrams, we used morpheme pairs (two consecutive morphemes) with a percentage of total morphological bigram total of 0.05% or more as features. This is an input vector by using a dummy variable and setting its value to 1 when a morpheme or a morpheme pair as a feature exists in the statement, and setting the value to 0 if it does not exist.

N gram is a co-occurrence relationship (collocation) of a language unit in which a certain language unit (character, morpheme, etc.) in the document occurs adjacent to each other such as two linguistic units, three linguistic units (generally N languages unit). It can be thought of as representing a kind of feature of the document. The N grams of two languages and three languages are called bigrams and trigrams, respectively. Also, the histogram (frequency distribution) of a single language unit is called unigram.

N grams when language units are morphemes are called morpheme N grams. Again, as N increases, it becomes enormous data, which makes management difficult. Therefore, bigrams and trigrams are mainly used.

When M is the total number of linguistic units (e.g., characters), the total number of possible combinations of consecutive N units is M^N. In the method of preparing a combination table of M^N, the number of occurrences (frequency) is checked from the text. In this case, the computation of N grams takes time with increasing N, and as a result, the data become extremely huge. It is extremely difficult to store the resulting data created by this method.

All morphemes are always given the attribute of a part of speech (except for words that are not in the dictionary, that is, unknown words). In most cases, the part of speech is not enough to express the meaning of words. For example, it is understood that the verb expresses "some kind of action", but to know the specific meaning, it is not enough with the part of speech information alone.

Then, how about using words themselves as meaning units? A word whose meaning is like another word is called a synonym. We think that the set of synonyms represents a certain meaning (or concept). It is unavoidable to use it as a unit for expressing the meaning of a word (or morpheme) that decomposes a word, since it is only a word to tell others what he is thinking to others.

However, of course, that might be a problem. It may be more convenient for you to be able to define similarities such as similar, not similar, very similar, slightly similar, and not very similar. For example, how much the meaning of "writing letters" and "speaking words" is similar, which is more similar to "drinking water" "eating fruit" or "reading a book", etc.

In other words, it is necessary to consider the question of whether the meaning of words whose classification is too rough in parts of speech is not expressed as a set of words themselves, but whether it can be expressed in a manner of high abstraction to the extent that similarity can be considered.

Recently, machine learning has proposed a way to treat words as vectors rather than letters. Context information is used to create that vector. In other words, it characterizes the meaning of the target word by another word around the word where the word appears. Such a vector is called a distributed representation of a word. In the case of characters, I care about whether or not they match, but if you have a vector, you can calculate the distance between vectors and handle it numerically as similarity. By introducing an indicator of distance to what is difficult to capture the meaning of words, it means that we can handle things more flexibly. Calculation increases, but computers will get faster over time, so you will not have to worry about this indefinitely.

Let's adopt the machine learning method. First, let's make a word vector so that it can be applied to the learner. There is a method called one-hot that uses a dictionary. Figure 2.11 shows the mechanism.

As shown in this figure, numbers are added in the order of posting of dictionaries, a vector matching the size of the dictionary is created, and the value of the dimension corresponding to the index of a certain word becomes 1, and everything else becomes 0. Now you can create a one-to-one vector for every word. It is called a one-hot vector.

Vectors made by one-hot are not suitable for calculating similarity because they do not use context information at all, such as co-occurrence relations. So, we devise the following ingenuity. This is a method known as word2vec, and there is not much theoretical background, but since it can be easily implemented and its effectiveness is confirmed to a certain extent, it is one example of vectorization.

Word2vec is a method of expressing all words by 200-dimensional vectors (number of dimensions can be changed), and it is possible to add and subtract words by using vector representation (Mikolov et al. 2013a, b). For example, if you enter "Tokyo—Japan + France" the answer will be "Paris" and if you enter "King—Male

2.7 Natural Language Processing for Deep Understanding of Discussion

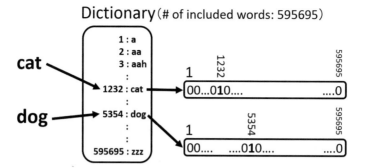

Fig. 2.11 One-hot word vectors

Fig. 2.12 Skipgram

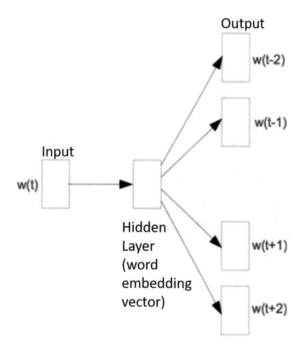

+ Female" the answer will be "Queen". This allows you to create a vector from a word by entering a corpus (actual document training data) into the neural network called skipgram shown in Fig. 2.12 and learning it.

A skipgram takes as input $w(1),\ldots, w(T)$ ($w(i)$ is a vectorized word with one-hot; T is the total number of words appearing in the document). Skipgram is the neural network for predicting words appearing before and after of the input $w(t)$ (t is a position of occurrence in the document). We use the stochastic gradient descent method to learn using the following formula:

$$\arg\max_{P} \frac{1}{T} \sum_{t=1}^{T} \sum_{-c \leq j \leq c, j \neq 0} \log p(w_{t+j}|w_t)$$

$$p(w_O|w_I) = \frac{\exp(v_{w_O}^T v_{w_I})}{\sum_{w \in W} \exp(v_w^T v_{w_I})}$$

Here, C is called the context size; it indicates how many words before and after the input are predicted, set to about 5. w_O is the output word, and w_I is the input word. In addition, v_w is a vector representing the word w, and it is assumed that it is calculated considering the regularity of the context in which the word appears. The dimension of v_w is set to about 200.

The result of the experiment is that the vector v_w representing each word has the following properties. The first is that the cosine similarity (a value obtained by dividing the inner product of two vectors by the product of the lengths of both vectors) between vectors of words that are likely to appear in the same context, such as "mathematics" and "physics", is larger than the cosine similarity between vectors of other words such as "mathematics" and "cooking". This can be thought of as approximating the semantic similarity between words as cosine similarity expresses the similarity between vectors in this case.

The other is a property called linear regularity between words. For example, as shown on the left side of Fig. 2.13, the cosine similarity between the difference vector between a vector representing "woman" and a vector representing "male" and the difference vector between the vector of "aunt" and the vector of "uncle", or the difference vector between the vector of "queen" and the vector of "king" was very high. This predicts that the semantic commonality between these words is high. Also, the right side of Fig. 2.6 shows that the relationship between singular and plural forms of words is common among words.

The third property is that the vector reflects the "degree" of the meaning of the word. For example, as shown in Fig. 2.14, taking an average vector of good and best vectors, there is better in a word with the highest cosine similarity (in other words, better is in the middle of good and best). Likewise, it turns out that unhappy

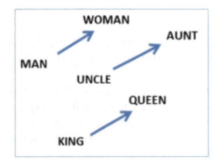
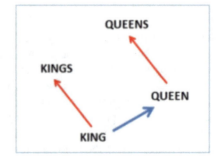

Fig. 2.13 Linear regularity

2.7 Natural Language Processing for Deep Understanding of Discussion

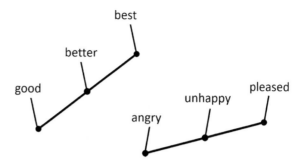

Fig. 2.14 Degree of meaning of words

is roughly in between pleased and angry, which can be thought of as reflecting the degree of a certain meaning of the word.

The last property of a word vector is language independence. Even if the language (such as English or Japanese) differs or the dictionary for which one-hot is calculated differs depending on sufficiently large training data (collection of natural language documents), similar regularity was obtained. This will need to be verified in various languages (the author of the original paper confirmed in English and Spanish), but it will be necessary to wait for sufficient data to gather for learning.

In this way, the mechanism of handling the meaning of a word on a computer is greatly developed by machine learning. From now on, more appropriate vectorization method will be devised, I think that subtle nuances that were difficult to handle with traditional systems are handled well and human intention will be recognized more deeply.

2.8 Natural Language Processing for Discussion Structure Analysis

There is a context in the discussion as well, so if you use it well, you can handle the meaning of the statement more accurately. In other words, the statement has another statement as a trigger for that, and it is related to that statement. Regarding the relationship between the statements, as already mentioned, the discussion concerning a topic is a tree structure, and that node is each statement. We apply natural language processing for this structured data of discussion.

As a result, you can see, for example, the following:

Whether someone's statements were appropriate in that context.
Whether a statement by a person still contains content to be considered.
Whether someone's statements were comprehensive and were good statements.

Based on the data obtained by discussion mining, we have constructed a machine learning model that classifies statements from the viewpoint of newly introducing

42   2 Discussion Data Analytics

topics or continuing previous topics. This allows you to classify speech recognition results in real time and segment them into several topics.

Features used for machine learning model construction are as follows:

(1) Whether or not the displayed slide image matches between a certain statement and the preceding utterance.
(2) Whether the two previous speeches are the same as the speaker.
(3) How long has passed since the last statement.
(4) Similarity with the preceding utterance (cosine similarity between vectors when speaking is represented as a vector expression).
(5) Morpheme unigram and bigram.
(6) Whether the slide was changed during speaking.
(7) Entity Grid (described later).
(8) Length of sentence included in statement.
(9) Time elapsed when pointed in the slide with the discussion commander.
(10) Whether or not it coincides with the pointed object pointed to by the immediately preceding utterance when using the pointer.
(11) Number of sentences included in the statement.
(12) Whether the reference includes a referring expression (such as a pronoun).

Entity Grid is a model based on the hypothesis that there is regularity in the distribution of appearance of the elements described in the consistent text, letting the sentences in the line correspond to the elements in the sentence (Barzilay and Lapata 2008). It is expressed as a matrix. For each term, the elements of the sentence correspond to what is called the syntactic role in that sentence of that sentence. Specifically, there are four types: Subject (S), Object (O), Other (X), and Not Emerging (–).

For example, an Entity Grid like Fig. 2.16 is created from the four sentences shown in Fig. 2.15.

Here, we decided to put "BELL INDUSTRIES Inc." and "Bell" of s1 and s4, respectively, into the same column named "BELL" as we concluded that they refer to the same thing through a co-reference analysis method. Both are the subject of the sentence, so the syntax role is S. Even when referring to the same target using pronouns or the like, make similar analyzes so that they are in the same column.

In order to use this Entity Grid as a feature vector, we investigate the change of syntactic role of a noun in two consecutive sentences and calculate transition probability. The transition probability of a syntactic role is calculated as the probability of occurrence of that transition in the entire transition. For example, in the example of the Entity Grid of the sentence of Fig. 2.15 (Fig. 2.16), since "BELL" is S in s1 and does not appear in the next sentence s2, we consider it as –, which is a transition of

s_1 [BELL INDUSTRIES Inc.]$_S$ increased [its quarterly]$_O$ to [10 cents]$_X$ from [seven cents]$_X$ [a share]$_X$.
s_2 [The new rate]$_S$ will be payable [Feb. 15]$_X$.
s_3 [A record date]$_S$ hasn't been set.
s_4 [Bell]$_S$, based in [Los Angeles]$_X$, makes and distributes [electronic, computer and building products]$_O$.

Fig. 2.15 Example text (from Paper Barzilay and Lapata (2008)

2.8 Natural Language Processing for Discussion Structure Analysis

Fig. 2.16 Entity Grid (from Paper Barzilay and Lapata 2008)

	BELL	quarterly	10 cents	seven cents	share	rate	Feb. 15	date	Los Angeles	products
s_1	S	O	X	X	X	-	-	-	-	-
s_2	-	-	-	-	-	S	X	-	-	-
s_3	-	-	-	-	-	-	-	S	-	-
s_4	S	-	-	-	-	-	-	-	X	O

$S \rightarrow -$. Transitions of syntactic roles in such two sentences are all 10 (the number of elements) $\times$ 3 (the number of transitions in successive sentences, i.e., s1–s2, s2–s3, s3–s4) = 30 times. The number of occurrences of the transition $S \rightarrow -$ is three times since "BELL" has s1 to s2 once, "rate" is s2 to s3 once, and "date" is s3 to s4 once. That is, the transition probability of $S \rightarrow -$ is 3/30 = 0.1.

In order to obtain the transition probability of this syntactic role between two consecutive utterances, we focused on Japanese case particle "ha". That is because the noun phrase before "ha" often represents the subject of the sentence. So, with respect to all the sentences included in the utterance, if the noun phrase immediately preceding the case particle "ha" is gathered and the existence of the noun phrase or the noun phrase that points to the same is confirmed in the next statement, we examined the syntactic role of a noun phrase and calculated its transition probability. The total number of transitions was analyzed and counted for the entire discussion content of the same meeting.

Machine learning was carried out by logistic regression analysis, and the training data used the minutes accumulated from discussion mining described earlier.

The F-measure when experimenting with the test data is also listed in Fig. 2.17. The F-measure to the right of the feature in the table represents the value when the feature was not used for learning. This indicates that statements were classified with sufficient precision.

We then aimed at extracting more advanced information from the minutes. It was necessary to analyze the language content of statements and materials in detail to achieve that purpose.

By the way, it can be said that inconsistent statements in the discussion are statements that describe topics that are different from topics up to that point. So, consider how to categorize follow-up statements as comments deviating from topics or not. Logistic regression analysis described earlier is used for classification. In this case, we calculate the probability value that the topic is diverted and use this value for the consistency evaluation of the statement. For this purpose, in addition to the linguistic features obtained from the minutes, we use the meta-information given to the minutes. The features used in this method are shown below:

Feature	F-measure
All features used	0.9555
Relationship between statement and presentation material	0.7611
Whether the speaker of statement was the same as that of two previous statements	0.9101
Elapsed time since previous statement	0.9434
Cosine similarity between statement and previous statement	0.9519
Morpheme unigram and bigram	0.9525
Whether the slide change was made while speaking	0.9537
Entity grid	0.9543
Length of sentence	0.9555
Duration of use of pen	0.9555
Region overlap of pen drawing	0.9555
Number of sentences in statement	0.9555
Presence of reference expression (e.g., pronouns) in statement	0.9555

Fig. 2.17 Features for classification of statements

(1) Features based on linguistic features of text:

- Relevance to parent statement.
- Single sentence or compound statement.
- Number of characters of statements.
- Morpheme unigram and morpheme bigram.
- Presence of subject and referring word.
- Entity Grid.

(2) Features based on meta-information attached to the minutes:

- Whether the speaker is a student or not, whether it is a presenter or not.
- Whether the speaker of the parent statement is the presenter.
- Presence of marking/agreement/disagreement buttons.
- Depth from the root in the discussion tree structure.
- Whether or not the visual referent of the parent statement matches that of the target statement (detailed later).
- Presence or absence of slide operation during speaking.
- Time for reservation of speaking.
- Presence or absence of different statements in time series between the parent statement and the target statement.
- Alternation of questioner.

2.8 Natural Language Processing for Discussion Structure Analysis

For morphemes and morpheme pairs that appear during speech, the number of occurrences of nouns, verbs, adjectives, auxiliary verbs, and morpheme pairs is calculated by preliminary survey as in the analysis of the task statements described above, we used those exceeding a certain value for the feature. Also, since there is a report that Entity Grid is effective for evaluating text consistency (Barzilay and Lapata 2008), it is directly related to topic transformation among the syntactic role of the Entity Grid. We focused on only the transition of the theme considered as a transition probability and used it for the feature.

The alternation of the questioner, which is the last feature, as a matter of whether or not the questioner is different from the preceding statement pair when considering participant's question and presenter's response as a statement pair.

We implemented the above method and conducted experiments on discrimination of inconsistent statements. As a dataset, we used 53 min (discussion content) of seminar in our laboratory (number of statements: 3553). However, since the start-up statements are not subject to this case, the follow-up statements (the number of statements: 2490 cases) are subject to discrimination. As correct answer data (teacher signal), we decided manually whether a certain statement lacked consistency, and gave that attribute. 202 consecutive statements were determined to be lacking in consistency.

In order to evaluate the proposed method, the case where learning was carried out without using features based on the meta-information of the minutes was taken as a comparative method. For the evaluation, we used the precision and recall described above, the F-measure, and carried out the tenfold cross-validation.

The results of this experiment are shown in Fig. 2.18. The results of the consistency judgment by the method we proposed are higher than the case where the feature information given to the minutes is not used; an advantage was confirmed in the precision, the recall, and the F-measure.

In addition, when learning by removing each feature by the meta-information of the minutes, the precision, the recall, and the F-measure declined in all the features, and the effectiveness of the used feature was confirmed. Figure 2.19 shows the results of the top five cases where the F-measure drops greatly.

	Precision	Recall	F-measure
Proposed method	0.269	0.534	0.358
Comparative method	0.117	0.129	0.123

Fig. 2.18 Experiment results of consistency checking

Removed feature	Precision	Recall	F-measure
Whether speaker is presenter	0.255	0.494	0.337
Presence of marking/agreement/disagreement buttons	0.251	0.522	0.341
Depth from root in tree structure (i.e., discussion chunk)	0.253	0.522	0.341
Whether visual referent of parent statement matches that of target statement	0.259	0.506	0.342
Alternating of questioners	0.255	0.534	0.345

Fig. 2.19 Feature contribution to learned model

2.9 Correctness of Discussion Structure

Each statement in a discussion generally contains several sentences. Therefore, when considering the structure of the discussion, it would be appropriate to think separately into the structure within the statement and the structure between the statements. The structure in the statement consists of the internal structure of the sentence contained in the statement and the relation between the sentences.

Regarding the internal structure of a sentence, there is another analysis method for the relationship between sentences. The structure that spans multiple sentences is called a discourse structure, but there is no decisive way to analyze how to analyze it. In one example, we use a conjunction connecting sentences as a clue. If it is connected with "and" like "it is A, and it is B", the two sentences are in a relationship of adjacency, that is, the later sentence supplements the information of the previous sentence. It is content which is semantically contradictory. In addition, "it is A, but it is B", the two sentences are inverse relations; there is a semantically contradictory content in the sentence to be expected from the previous sentence. In addition to this, there are parallel, reason, assumption, limitation, time lapse, topic expansion, and topic convergence among others in relation to such sentences.

If such a conjunction is not described, it is necessary to predict the relationship between the sentences by using the contents of the two sentences (or the sentences before and after them). As for this also, a machine learning method can be considered. However, in this case, unlike syntactic analysis, it depends heavily on semantic content, so it is better to use features reflecting the meaning in the word context, such as the vector expression by word2vec described earlier.

Well, next is the relation between the statements, as in the first part of this chapter, we classify the statement with two tags "start-up" and "follow-up", and in the case of "follow-up" statement, it is structured in a way to link to the original (parent)

2.9 Correctness of Discussion Structure

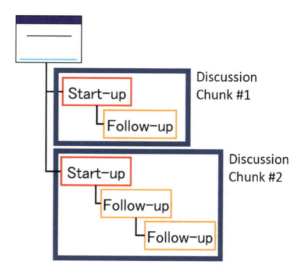

Fig. 2.20 Discussion chunk

statement. We use a discussion structure tagged and linked by the speakers as a correct answer, but in the first place, we need to verify how valid these tags and links are.

By the way, in the discussion mining (DM) system we developed, we are recording and analyzing all the contents of the discussion on presentations using presentation slides created with Microsoft PowerPoint. During the meeting, with the presentation slide displayed by the presenter, when the participant makes a start-up statement and the presenter or another participant makes a follow-up statement, a series of steps until the next start-up statement is made. We call this structure a discussion chunk. An example is shown in Fig. 2.20. A discussion chunk is considered a discussion on one topic.

In the DM system, there is a part depending on participant's judgment when creating discussion content, so there is a possibility that human error such as missing information or inputting wrong information may occur. In addition, since we developed a minute input tool that does not hinder discussion as much as possible, there is a possibility that the intention of the speakers may not be fully reflected in the acquired metadata. Among the metadata acquired by the DM system, the tree structure of the discussion chunk and the text summary of the statement are greatly affected by human judgment, in particular. Whether or not the statement text correctly reflects the spoken content is thought to be able to be checked mechanically by using speech recognition which is also described in the next chapter, but as for the structure, if a human does not give a correct answer, machine learning does not work well. So, its accuracy is a big issue.

Therefore, we conducted the following experiment to evaluate the validity of the tree structure of the discussion chunks (Nagao et al. 2005). In this experiment, we ranked the top 18 segments with many statements out of discussion chunks in discussion content created after 2007. The total number of statements included in the 18 chunks was 199 (of which 181 statements were follow-up statements), the average of the number of statements per chunk was 11.1, and the average of the number of speakers in the chunk was 4.6. In this experiment, the accuracy of data was evaluated by comparing the statement type and the tree structure generated by the DM system with the correct answer data created manually regarding the following items:

(1) Follow-up statements presenting new topics:
   As an empirical problem, it often happens that the topic gradually diverges as the discussion becomes longer. Therefore, as examining individual speech contents included in a discussion chunk with a large number of statements, there is a possibility that the contents and intent of the start-up statement as the root of that segment may include a follow-up statement that does not continue the topic. Therefore, by examining the number of follow-up statements presenting new topics in the chunk, we verified the validity of follow-up statements.

(2) Follow-up statements with incorrect parent statements:
   The statement reservation function of the DM system mentioned earlier gives link information between the statements based on the input time of a reservation when the target statement is being made. Therefore, if the user makes a reservation after the statement has ended, the correct link information may not be given. For that reason, we evaluated the validity of the statement reservation function by the number of follow-up statements of incorrect parent statements (previous statements directly linked with the follow-up statements).

(3) Follow-up statements with multiple links:
   In the current statement reservation function, the number of parent statements is limited to one, but when making statements that summarize the ongoing discussion in the chunk, more than one statement may be referred to. Therefore, in order to check whether the link information acquired by the statement reservation function is sufficient, we examined the number of follow-up statements with multiple links.

Figure 2.21 shows an example of a comparison between the discussion structure generated by the system and correct answer data. In this example, the statements corresponding to (1) above is 6, the statements corresponding to (2) are two statements 3 and 5, the statement corresponding to (3) is the statement 4 that has links to 1 and to 3.

Correct answer data were created by the following procedure. First of all, in order to confirm whether the contents and intention of the start-up statement serving as the root of the discussion chunk are correctly reflected in the subsequent follow-up statements, we asked the speaker of the start-up statement of the target chunk to select statements that deviated from the intention of the start-up statement. Next, we asked several university students that participated in the target discussion to make the correct answer for the above items (1), (2), and (3) while viewing the contents with the discussion browser mentioned earlier. Finally, we reviewed the students' answers and decided the final correct answer data.

Figure 2.22 shows the results of comparing the structure of the discussion generated by the DM system with the correct answer data.

2.9 Correctness of Discussion Structure

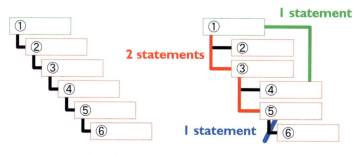

Fig. 2.21 Comparison of discussion chunks

ID	Statements	Speakers	Statements with new topic	Statements with incorrect parent	Statements with multiple links
1	14	6	1	2	2
2	13	5	1	0	3
3	13	5	1	0	3
4	10	4	2	0	0
5	13	4	2	0	2
6	8	4	1	0	2
7	7	4	1	0	1
8	6	4	1	1	1
9	17	5	1	0	1
10	11	4	1	0	2
11	11	4	1	0	0
12	5	3	0	0	0
13	16	5	6	0	0
14	13	5	1	1	3
15	14	4	1	0	1
16	11	7	2	0	0
17	9	5	1	0	1
18	8	5	0	0	1

Fig. 2.22 Experimental results

The consideration obtained by this experiment is as follows:

(1) Tendency of follow-up statements presenting new topics:
From Fig. 2.16, there were some follow-up statements presenting new topics in almost all discussion chunks. In other words, we can see that there are multiple topics in one chunk. The most frequent follow-up statement presenting a new topic was a statement that presents a topic that was inspired from the discussion up to the present. For example, until the point just before, the participant was discussing "can you split semantic units?", the statement that cuts out a topic from another viewpoint "can you unite semantic units in reverse?" was made. With the current statement types, we cannot divide the discussion in this flexible way. The longer the length of one segment is, the longer the time required for browsing becomes. Therefore, in order to realize efficient content browsing, finer segment information is acquired by combining a technique different from the statement types.
If the speaker of the start-up statement can judge whether a follow-up statement is not related to the content or intention of his/her statement, it is thought that finer segment information can be acquired.

(2) Tendency of follow-up statements with incorrect parent statements:
Four follow-up statements with incorrect parent statements were confirmed. It was confirmed that it is equivalent to 2.2% in all follow-up statement (181 statements). From this, it can be thought that the link information of the statement acquired by the statement reservation function is valid. Also, when examining the corresponding statements in detail, it turned out that it was caused by delayed timing of speaking. The reservation function gives link information to the statement that was being done at the input time of a reservation, but when the statement is not made at the time of reservation, the reservation function is designed to give link information to the last one (of course, this rule is not applied to the start-up statements). Therefore, the system cannot attach link information to the statements made before the last statement. As a solution to this problem, there is a method of specifying the statements to be targeted when the speaker speaks. For that reason, we added a function to change the linked statement (parent statement) to the discussion commander that is a device held by each participant.

(3) Tendency of follow-up statements with multiple links:
As you can see from Fig. 2.22, many follow-up statements with multiple links existed. As checking the individual statements, there are a lot of statements to express opinions and summaries based on statements and discussions until then, such as "to arrange because discussions are complicated" or "even though you were saying …". In addition, many chunks that do not have follow-up statements with multiple links repeatedly asked questions and requested. From this, it can be said that "the statements with more links are more important in the discussion," "the discussion is more active as the number of follow-up statements with multiple links increases," the link information can be thought of as an index for the importance and activity level of discussion.

It is impossible to acquire multiple link information only with the current statement types and the statement reservation function, but it can be dealt with by using the discussion commander to add the statement to be referred to as well as the correction of the parent statement. In addition to the method of inputting in real time, it is possible to consider how to input it after the meeting. For example, by realizing a mechanism for describing ideas by quoting multiple statements that are considered important, the system can give link information that shows relevance between quoted statements at the same time.

2.10 Structuring Discussion with Pointing Information

As described in the previous section, the structuring of discussion chunks in our discussion mining system is problematic. For example, more than one topic may be included in a discussion chunk. Experiments showed that the speakers who contributed to a discussion chunk were not always talking about the same topic. As the discussion continued, several opinions from various viewpoints tended to be given. We attribute this to two main reasons. The first is that a new topic was introduced to enrich the ongoing discussion. It would be unusual for all the remarks made during a discussion to be about a single topic; it is more common for supplemental topics to be introduced during a discussion. The other reason is that the statement types (start-up and follow-up) used were not sufficient to classify all the statements in the discussions. When a different topic was introduced in a single statement of a discussion chunk, the system could not create a sufficiently detailed semantic structure.

When a user reviews the discussions of previous meetings by using the discussion browser, the statements made during the meetings are numerous, and even if metadata is available to help in the search effort, it may take a long time for him or her to select and read statements of interest. Therefore, further structuring focusing on topic segmentation as well as structuring on the basis of "start-up" and "follow-up" is required.

In our research, we focus on pointing and "referring to" behaviors during meetings. Speakers usually refer to something when making a statement, e.g., "this opinion is based on the previous comment" or "this is about this part of the slide (while pointing to an image or text in the slide)." We assume that a statement with a reference to an object in a slide is strongly related to the topic corresponding to the object. We also assume that two statements during which the speakers point to the same object are about the same topic. Therefore, we concluded that acquiring and recording information about pointing to objects in a slide would facilitate topic segmentation and lead to more precise semantic structuring of discussions. We call a pointed object in a presentation material a visual referent.

We thus developed a system for pointing to and selecting objects in slides that uses the discussion commander mentioned earlier and created a mechanism for acquiring and recording pointing information related to participants' statements.

This system can also extract any part of the figure in a slide and refer to it. In addition, selected or extracted image objects can be moved and magnified by using the discussion commander.

The system also gathers various types of metadata about pointing such as identifiers of objects and slides, starting and ending times of pointing, IDs of participants who are pointing at objects, and IDs of statements during which objects are referred to. The significant features of our pointing system are data correctness and clear visualization of "pointed to" objects. In a previous study (Nakano et al. 2006), the pointing target was estimated automatically by using a conventional laser pointer and visual processing. However, it was difficult to accurately acquire the pointing target. In this study, the pointing target is defined using information on objects in slides generated by Microsoft PowerPoint. This enables the attainment of references that correctly reflect the speaker's aim.

In our system, discussions are structured on the basis of metadata acquired by using our pointing system. We hypothesize that statements with the same pointing target are on the same topic with a high probability. Similarly, it is highly probable that the topic changes when the pointing target changes. Therefore, when a statement in a discussion chunk has the same pointing target as the previous statement, the system provides new link information between the two statements. This enables tightly connected statements to be distinguished from normally connected statements in a discussion chunk. This reference-sharing link information, which adds new meaning to the discussion structure, contributes to structuring discussions in more detail. Of course, there are still statements on the same topic without any visual referents. Since the reference is not visualized, it is difficult to define a target topic. Therefore, this type of topic sharing is removed from the scope of this research.

We evaluated our improved structuring method by using data from actual meetings. First, we verified our hypothesis that a statement with the same pointing target as the previous statement is, with a high probability, continuing the discussion of the current topic. Then we estimated the validity of structuring using pointing information by evaluating the validity of this hypothesis.

The discussion commander enables a user to point at and select objects shown on the screen. The user interface for selecting screen objects supports two methods as shown in Fig. 2.23.

1. Sectional detail drawing (left of Fig. 2.18): a rectangular area on the screen and chosen objects in the slide touched its rectangular range by using drag-type operations of a pointer cursor on the screen.
2. Underline drawing (right of Fig. 2.18): text in a slide is pointed at and selected by underlining it.

The references to various targets such as image objects and text portions in the slide are made possible by using the method appropriate for the target. These methods can also be used to acquire rich metadata about pointing.

Several types of objects are selectable: itemized text, grouped diagrams, images formatted in JPEG, BMP, AutoShapes (illustrations such as polygons and ovals that can be used by Microsoft Office applications), and tables and graphs.

2.10 Structuring Discussion with Pointing Information

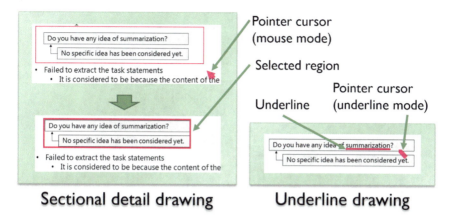

Fig. 2.23 Two methods for pointing in presentation slide

Four functions are provided that enable correct pointing to these objects using the sectional detail drawing method:

1. Temporary selection: a reference to an object is temporarily reserved.
2. Persistent visualization: the currently selected object and by whom are visualized.
3. Independentization: the function by which it is cut down as a target of pointing which doesn't depend on slide layout for enabling its movement and zooming.
4. Extraction: any subarea in the slide can be made independent.

The pointing information is used to structure discussions, i.e., to provide new link data between those statements referring to the same object, as shown in Fig. 2.24. Statements 3–6 in the figure are assumed to refer to the same object. Therefore, links are added to the discussion tree so as to make a complete graph consisting of statements 3–6. Statements referring to the same object (i.e., statements about the same topic) are connected by such links. By using these links, the system can distinguish a set of statements that are strongly connected with each other from a discussion chunk.

The speakers of statements 3–6 should be able to see that pointing to the same object created these links. However, they may find it difficult to ascertain whether their current statements are related to the "pointed at" object during the meeting. This is because they are concentrating on speaking when they are making statements even if the pointing target is visualized and can easily be confirmed. Of course, a speaker may point to an object and make a question consciously about it. As the discussion continues, the speaker may tend to forget to decide whether his or her statements are related to the "pointed at" object at that time.

We thus included in our system a confirmation dialogue: "is the present pointing target related to your statement?", and the next speaker should answer "Yes" or "No". A speaker operates his or her discussion commander and decides whether the "pointed at" object is related to his or her statement content. He or she selects "Yes"

Fig. 2.24 Links to shared visual referent

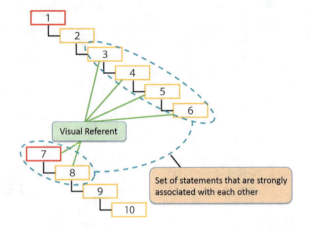

to continue referring to it or selects "No" to cancel the reference. This function can prompt the speaker to consider whether his or her statements should continue to refer to the object pointed at by the previous speaker when beginning a statement.

It is possible to add links that reflect the speaker's aim more accurately by using this function. The system can thus structure discussion chunks so that the speaker's aim is reflected correctly and thereby support browsing of the discussion contents more efficiently.

As described above, our proposed method for identifying statements in a discussion chunk with strong connections is based on the assumption that statements referring to the same pointing target are very likely about the same topic. However, this assumption was unverified. We thus conducted an experiment to confirm whether the assumed correlation exists and whether the derived links are correct. Specifically, we verified the hypothesis that "a follow-up statement that refers to the same pointing target as its parent statement is about the same topic, and it is rare that a new topic is introduced in such a statement."

We used two types of data in our experiment to verify this hypothesis: (1) pointing data captured by our discussion mining system, and (2) questionnaire data about the relationships between follow-up statements and topics acquired using the discussion browser. We collected the first type of data by using the discussion commanders of the participants in three laboratory seminars. The participants in each seminar were the professor and three or four students (ten students in total). Each student made a presentation with slides.

The participants were asked to make statements and to use their discussion commanders to point to objects on the slides during their statements. To point to paragraph text or the entire slide, they used the sectional detail drawing function. To point to a part of the text, they used the underline drawing function. To point to a particular region in the image, they used the independentization function. We removed the data corresponding to when they pointed to unintended objects by mistake.

2.10 Structuring Discussion with Pointing Information

We classified their statements into three types: statements with no related pointing, statements accompanied by a status change in pointing (start or end pointing), and statements accompanied by no status change in pointing (pointing at an already selected object, i.e., inheritance).

After each seminar, the participants were asked to complete a questionnaire on the discussion browser. However, we first had the participant who inputted the meeting contents by hand check to see if there was anything missing in the minute text since the minute text was an important source for the participants to complete the questionnaire.

The questionnaire consisted of three types of questions: (1) ones asking about the relationship between the statement of interest and the preceding start-up statement (the first statement of the discussion chunk in which the statement of interest was included); (2) ones asking about the relationship between the statement of interest and its parent statement; and (3) ones asking about the number of topics included in the statement of interest.

Our analysis of the questionnaire data revealed that, when a participant was referring to the same pointing target as the previous speaker, there was a high probability that he or she was speaking about the same topic. Our finding that there were few cases in which more than one topic was introduced in a statement validates our hypothesis that "speakers who make statements while referring to the same pointing target are very likely to be speaking about the same topic in contrast to statements without a shared reference, and it is rare that a new topic is introduced in such a statement".

Our experiment revealed that some follow-up statements were about a topic differing from that of the start-up statement. The discussion may thus become unsettled and then be abandoned because the participants do not know whether the discussion about the previous topic reached a conclusion. We may be able to develop a mechanism that can automatically identify such unsolved topics and suggest to the participants that they discuss them again.

We aim to achieve more semantic structuring of discussions by creating connections between statements including linguistic co-references and anaphora.

References

R. Barzilay, M. Lapata, Modeling local coherence: an entity-based approach. Comput. Linguist. **34**(1), 1–34 (2008)

K. Georgala, A. Kosmopoulos, and G. Paliouras, Spam Filtering: An Active Learning Approach using Incremental Clustering, in *Proceedings of the 4th International Conference on Web Intelligence, Mining and Semantics*, No. 23 (2014)

D. D. Lewis, W. A. Gale, A Sequential Algorithm for Training Text Classifiers, in *Proceedings of the ACM SIGIR Conference on Research and Development in Information Retrieval*, pp. 3–12 (ACM/Springer, 1994)

A. Liu, L. Reyzin, B. D. Ziebart, Shift-Pessimistic Active Learning using Robust Bias-Aware Prediction, in *Proceedings of the AAAI Conference on Artificial Intelligence* (2015)

T. Mikolov, I. Sutskever, K. Chen, G. Corrado, J. Dean, Distributed Representations of Words and Phrases and their Compositionality, in *Advances in Neural Information Processing Systems*, ed by Burges, C.J.C., Bottou, L., Welling, M., Ghahramani, Z., Weinberger, K.Q., pp. 3111–3119 (2013)

T. Mikolov, K. Chen, G. Corrado, J. Dean, Efficient Estimation of Word Representations in Vector Space. *arXiv preprint*, arXiv:1301.3781 (2013)

K. Nagao, K. Hasida, Automatic Text Summarization Based on the Global Document Annotation, in *Proceedings of the Seventeenth International Conference on Computational Linguistics (COLING-98)*, pp. 917–921 (1998)

K. Nagao, K. Inoue, N. Morita, S. Matsubara, Automatic Extraction of Task Statements from Structured Meeting Content, in *Proceedings of the 7th International Conference on Knowledge Discovery and Information Retrieval (KDIR 2015)* (2015)

K. Nagao, K. Kaji, D. Yamamoto, H. Tomobe, Discussion Mining: Annotation-Based Knowledge Discovery from Real World Activities, in *Advances in Multimedia Information Processing—PCM 2004, LNCS*, Vol. 3331, pp. 522–531 (Springer, 2005)

W. Nakano, T. Kobayashi, Y. Katsuyama, S. Naoi, H. Yokota, Treatment of Laser Pointer and Speech Information in Lecture Scene Retrieval, in *Proceedings of the 8th IEEE International Symposium on Multimedia 2006*, pp. 927–932 (2006)

T. Nishida, Conversation quantization for conversational knowledge process. Int. J. Comput. Sci. Eng. **3**(2), 134–144 (2007)

N. Roy, A. Mccallum, Toward Optimal Active Learning through Monte Carlo Estimation of Error Reduction, in *Proceedings of the 18th International Conference on Machine Learning*, pp. 441–448 (ICML 2001)

B. Settles, M. Craven, An Analysis of Active Learning Strategies for Sequence Labeling Tasks, in *Proceedings of the Conference on Empirical Methods in Natural Language Processing* (Association for Computational Linguistics, 2008)

B. Settles, *Active Learning Literature Survey*, Computer Sciences Technical Report 1648, University of Wisconsin-Madison (2010)

H. Shimodaira, Improving Predictive Inference under Covariate Shift by Weighting the Log-Likelihood Function. J. Stat. Plann. Infer. **90**, 227–244 (2000)

M. Sugiyama, M. Kawanabe, *Machine Learning in Non-Stationary Environments: Introduction to Covariate Shift Adaptation* (The MIT Press, 2012)

T. Schultz, A. Waibel, M. Bett, F. Metze, Y. Pan, K. Ries, T. Schaaf, H. Soltau, W. Martin, H. Yu, K. Zechner, The ISL Meeting Room System, in *Proceedings of the Workshop on Hands-Free Speech Communication (HSC-2001)* (2001)

Chapter 3
Creative Meeting Support

Abstract A face-to-face meeting is one of the basic social activities; it is necessary to analyze it in order to understand human social interaction in detail. This research is a scientific analysis of human social interactions. We are researching a mechanism to promote innovation by supporting discussions based on the premise that innovations result from discussions. Ideas are created and developed during conversations in creative meetings like those in brainstorming. Ideas are also refined in the process of repeated discussions. In our previous research of discussion mining, we specifically collected various data on meetings (statements and their relationships, presentation materials such as slides, audio, and video, and participants' evaluations of statements). We developed a method to automatically extract important statements to be considered after the meetings by using the collected data. Actions such as investigations and implementations are performed in relation to these statements. Here, we present an idea that automatically extracted statements leading to innovations facilitate creative activities after meetings. Our research was aimed at deeply analyzing face-to-face meetings and supporting human creative activities by appropriately feeding back knowledge discovered in the meetings. We particularly analyzed the features of statements made during discussions. We developed a system called a "meeting recorder" for that purpose. The meeting recorder consists of a 360° panoramic video camera that records meetings in audio–visual scenes, a tablet application that allows users to browse meeting materials and add various notes to them with a stylus, speech recognition that identifies speakers and transcribes speech contents of all meeting participants, and a minute server that integrates all meeting-related information and creates the meeting minutes. We also developed a system that supports the activities after meetings called the "creative activity support system." This system supports users in quoting statements extracted from the minutes, in writing notes and reports, in creating activity plans, in managing schedules to accomplish tasks, and in evaluating other members' results within the group.

Keywords Meeting analytics · Meeting recorder · Creative activity support · PDCA cycle · Evaluation of creativity

© Springer Nature Singapore Pte Ltd. 2019
K. Nagao, *Artificial Intelligence Accelerates Human Learning*,
https://doi.org/10.1007/978-981-13-6175-3_3

3.1 Meeting Analytics

Along with the remarkable development of machine learning technology, data analytics, natural language processing, and pattern recognition technology, which I will describe later, have greatly improved. Of course, that did not solve all the problems of intelligence, but at least from several years ago, the work we can automate has increased considerably. Recent artificial intelligence introduces machine learning technology (or improves machine learning mechanism previously introduced) to various human-supported information systems which have been researched and developed for quite some time. The systems with recent artificial intelligence technology are a system that can become smarter by collecting data.

There is also a place where artificial intelligence can be active even in the area of meeting support. It is not a story of automating the meeting itself to make human beings unnecessary but to try to make artificial intelligence take over work of the surroundings so that human beings can fully demonstrate creativity and improve productivity. Even at the meeting, various data can be gathered by incorporating various ideas into the system, which can be used for machine learning. However, as explained in Chap. 2, in the case of supervised learning, a teacher signal (correct answer data) is necessary. Also, by using the active-learning method also described in Chap. 2, human beings can update their learning models by making appropriate feedback with little effort. The learning model here is a predictive probabilistic model that is useful for appropriately advancing the meeting, efficiently reviewing the contents of the discussion, and appropriately reflecting it in subsequent activities. Specifically, this model contributes to improving the accuracy of recognition and summarization and finding important issues more accurately.

The mechanism to promote the improvement of discussion ability will be described in the next chapter of this book, but here I will explain another meeting support system and its functions we developed.

For creative meeting support, we applied machine learning techniques for discussion activity data obtained using a discussion mining system described in Chap. 2 that structures the meaning of a meeting's content and records it. We call this discipline of analyzing meetings and automatically mining valuable information from them "meeting analytics."

As an important tool for meeting analytics, we developed a system called the meeting recorder that records and analyzes small-scale and face-to-face meetings in detail. It consisted of an Internet of Things device with a 360° all-around panoramic camera, tablets for all participants, and a server computer. This was aimed at achieving a simplified portable version of the discourse mining system and it was also aimed at fully automating the creation of the minutes.

The system records meeting scenarios with a panoramic camera and tracks participants' faces by assuming that there are people walking around in meetings. The system also simultaneously collects all the participants' voices with small wearable microphones. The speech recognition results and meeting materials are synchronously displayed on the tablets of participants. In addition, users can freely

3.1 Meeting Analytics

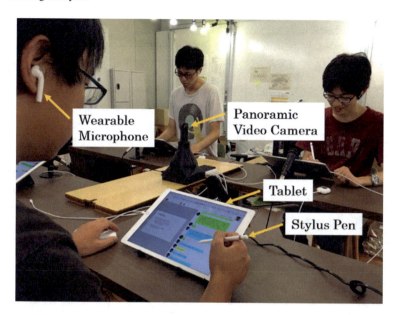

Fig. 3.1 Meeting recorder in action

draw and mark figures on documents with stylus pens. The stylus inputs are also shared on all the participants' tablets in real time.

There is a photograph of the meeting recorder we developed in use in Fig. 3.1.

The voices of meeting participants that are input for the wearable microphones are recorded on the meeting server together with the IDs of the speakers along with when their speech started and ended. These are simultaneously transcribed using speech recognition on the cloud and displayed on the tablets of all participants. The facial images of the speakers obtained from the panoramic camera are then also simultaneously displayed.

The material displayed on tablets are images of a file (mainly in PDF format) transmitted beforehand to the meeting server, which can be marked with a pen linked to the tablets. Materials and markings are synchronously displayed on the tablets of all participants, and the statements and the materials (including markings) are automatically associated and used for structuring the discussion, which will be described later.

The meeting recorder is a system developed to support human creative activities by analyzing more general meetings not being limited to seminars of university laboratories and by appropriately feeding back information (Nagao 2018).

In particular, this system can analyze the characteristics of utterances in a voice discussion and create discussion content like the discussion mining (DM) system described in Chap. 2.

The biggest difference between this system and the DM system is that it is portable. In the DM system, it was necessary to install a pan-tilt camera and a device called an IR array that sent infrared rays to the ceiling. As a result, meetings could only be held in certain rooms. In order to take a lot of data, it is necessary for everyone to be able to use it easily, so it was necessary to redesign to make installation easier.

The meeting recorder includes a panoramic camera that records meeting landscapes, a tablet application that allows users to browse slide materials and freely add them with a pen, a speech recognition system that identifies speakers and transcribes the contents of utterances. The tablet and the panoramic camera are connected to the minutes server by wireless network. In other words, if you install a server in the same facility, you can carry out a meeting using the system by carrying only the tablet and the camera. In addition, all the participants are wearing a small wearable (wearable) microphone on one ear. This is so that participants can record and recognize that person's voice without fail even if they walk around. This microphone is connected wirelessly to the tablet, voice is transmitted to the speech recognition cloud (e.g., Google or Amazon) in real time via the tablet, and the recognition result is collected in the minutes server.

On the minutes server, machine learning is used to estimate statements that change the topic in discussion. By doing so, the system can divide the minutes by topic, and consider the number of statements and the number of speakers in that topic, then the system can decide the importance of each topic.

If the discussion can be further structured, more sophisticated processing can be realized. In order to structure it, it is necessary to gather a lot of data. This system collects and analyzes various data including audio and video about meetings in the simplest possible way.

For example, as described in Chap. 2, we construct a tree structure using the relationship between statements and between statements and slide materials, and then perform spreading activation on it. Spreading activation is a mechanism that gives a high activation value to a node with a high degree of reference on the graph, and by using this activation value, it is possible to calculate the importance of the statements.

The system can select topics with high priority and statements with high importance and present them as a summary of the meeting minutes. Also, the system will be able to remind the participants to discover statements that describe important issues to be solved in the near future, such as the task statements mentioned in Chap. 2.

360° spherical panoramic cameras (or simply panoramic cameras) are now being sold for the general public. This is a camera that can shoot pictures and images of 360° around the camera. Because it is a mechanism with two fisheye lenses overlapped, a distorted image as shown in Fig. 3.2 is taken, so use PC to convert it to an image like a normal camera shot and display it. In Fig. 3.2, we also show the results of recognizing the persons and things in the image.

In the DM system, the face of the speaker was photographed using a pan-tilt camera (a camera that can be turned by PC) on the ceiling, but when the speaker walks around and moves to a place, the camera cannot follow. On the meeting recorder, we assume that there are people walking around while meeting and recording meeting scenes with a panoramic camera and tracking the participants' faces.

3.1 Meeting Analytics

Fig. 3.2 Image captured by panoramic camera

This is so that you can easily search the speaker's current video from the speech when browsing the recorded content later. By analyzing the position of the participant's face in the panoramic image, it is possible to find the image of the speaker at that time in the panorama from the speaker ID and the speaking start time.

Also, as mentioned above, participants all wearing a small microphone in their ears and participate in the conference, and the voice of all the participants is recorded separately. Then, as shown in Fig. 3.3, the speech recognition result and meeting materials such as presentation slides are displayed synchronously on the tablet of each participant. On the right side of the screen shown in Fig. 3.3, the speech recognition results are displayed in a chat log format. The user's own utterance is displayed right aligned, the other utterances are displayed left aligned, and color coded for each person. This color also matches the pen stroke color of the corresponding user. Meeting materials are displayed on the left side of the screen, so that writing can be done with a pen. The user can also use it like a whiteboard by making the background white, without displaying the document. At the top of the screen, a part of the utterance content that represents the topic at that time is displayed.

The minutes created by the meeting recorder include images, sounds, images, texts, and can be searched and viewed using a Web browser. The user interface of the minutes is shown in Fig. 3.4. Images shot with panoramic camera, slide image displayed on tablet marked with pen and recognized speech text are displayed. When viewing video, the speaker automatically scrolls left and right when the speaker changes and the speaker is displayed. The slide image and the speech content are associated with each other, and an ordered list of statements is displayed for each slide image. Also, based on the discussion structuring method described in the next section, the statements are segmented into topics and are displayed in a tree structure format.

62   3 Creative Meeting Support

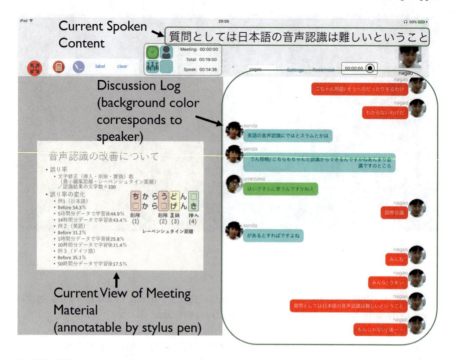

Fig. 3.3 Tablet interface of meeting recorder

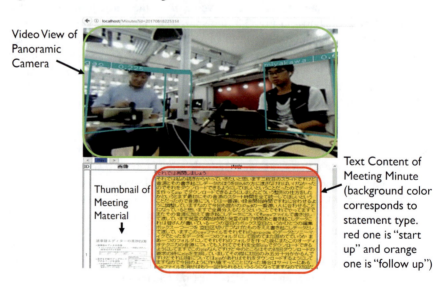

Fig. 3.4 Browser view of discussion content

3.2 Machine Learning for Structured Meeting Content

Although the online minute of the meeting recorder has the same function as the discussion browser mentioned in Chap. 2, the discussion does not constitute a tree structure at this point. That is because the meeting recorder is not designed to manually tag the statement types and to specify the parent statement in the case of follow-up statements in order to reduce the burden on participants. Therefore, we need to structure the discussion recorded by the meeting recorder based on the training data of the discussion content created and accumulated in the discussion mining system. This makes it possible to use various functions used in discussion mining. The functions include summarization of discussion content and discovery of task statements described in Chap. 2.

Regarding how to construct a tree structure, we begin by classifying the set of statements by topic. However, to understand what topics are, we need a fairly sophisticated semantic processing, so we modeled the tags ("start-up" and "follow-up") given by discussion mining and employed a machine learning technique as described in Chap. 2. If a statement that is supposed to be start-up type appears, the system considers that the topic has changed. In this way, since tagged statement data of discussion contents accumulated in the DM system can be used as it is as training data. Of course, using the data of the meeting recorder, if we perform active learning as mentioned in Chap. 2, we will be able to improve classification accuracy with minimal effort.

Assuming that tags of "start-up" and "follow-up" could be estimated for the statements, the next thing to do is to investigate which statement before a follow-up statement is related to the follow-up statement. In order to solve this problem, we apply a technique called conditional random field (CRF) (Qu and Liu 2012). This method gives attributes in a probabilistic way to each data included in the sequence data (the data arranged in order). As in speech recognition, the one-dimensional conditional random field is suitable for problems such as allocating phonemes to morphemes and assigning phonemes to divided waveforms. In addition, the two-dimensional conditional random field is suitable for the problem of estimating the dependency between data in series data. Therefore, in order to estimate the dependency between the statements and construct the tree structure of the discussion, we will use the two-dimensional CRF (2D CRF).

The 2D CRF has the structure as shown in Fig. 3.5. In this figure, the dependent element (i.e., follow-up statement) is called the source, and the depended element (i.e., parent statement) is called the target. The CRF is one of stochastic graphical models (graphical representations in which nodes are random variables) represented by undirected graphs (graphs in which nodes are connected by undirected links), and the link represents the dependency between random variables.

In Fig. 3.5, the case where the same statement has different parent statements is represented by nodes in the vertical direction, and the case where several follow-up statements depend on the same parent statement are represented by nodes in the horizontal direction. Like other machine learning methods, we estimate the parameter θ that maximizes the conditional probability of using the stochastic gradient descent

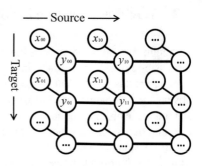

Fig. 3.5 Two-dimensional conditional random field

method. X is called the observation value (objective variable) and Y is called the latent variable (explanatory variable). Although details are omitted here, this mechanism makes it possible to structure the discussion recorded by the meeting recorder as a tree structure similar to the DM system.

In the case of content recorded with the meeting recorder, we regard the region marked with a pen stroke within the presentation material instead of a pointer as a visual referent (it is not necessarily a slide if it is a meeting material). Since the marking is shared by all the tablets, the user can select other people's marking area and make a statement about it, and the system can check the match of the visual referent between the statements.

This makes it possible to apply the model learned with discussion mining data to the data of the meeting recorder. Of course, depending on the content of the meeting, the prediction accuracy by the machine learning model may not be sufficient, so we need to update the machine learning model using the active-learning method described in Chap. 2. In order to do that, after the meeting, the system needs to get the user feedback which is the information equivalent to the teacher signal.

I will next explain how the system that supports the users' actions after the meeting and acquires data to improve the machine learning model at the same time.

3.3 Post-meeting Assistance to Support Creative Activities

We developed another system to extract important statements and task statements and to reflect them in subsequent activities to make effective use of the meeting results. This system consists of a function that involved writing notes and reports by quoting statements, a function that involves managing schedules concerning the execution of tasks, a function that involves disclosing and sharing notes and reports in a group, and a function that involves group members mutually evaluating reports. We call the system a "creative activity support system."

The creative activity support system is a system that guides subsequent activities from immediately after a meeting has been held based on the results obtained from discussions. As there is a possibility of forgetting content for a while after meetings, particularly for task statements, and there is a possibility of neglecting this, it is necessary to associate this with the activities as soon as possible. We implemented functions that users could easily remember to quote task statements, formulate action plans, reflect on schedules, and be appropriately reminded of the task execution plan for that reason.

It is not easy to steadily complete various tasks that arise in work and research. Indeed, there have been many arguments on how to manage tasks to achieve success, and many scheduling support systems, such as Google Calendar, have been developed to support this need.

Most conventional scheduling support systems have only been focused on managing the schedule of plans, and not on understanding how an established schedule is processed and what state it is in before proceeding to the next task. There is no support in implementing tasks that have causal relationships over the long term, such as setting guidelines. One major issue related to graduation and completion studies is addressed by carrying out individualized tasks that have been segmentalized in a long-term time series, so the plan–do–check–act (PDCA) cycle in business execution is recommended for education as well (Osone and Uota 2015). It is necessary to have a mechanism to support the smooth execution of a series of performance cycles of planning, executing, and evaluating tasks.

By using the creative activity support system, the users quote the statements extracted from the minutes, write notes and reports, form action plans and schedules them. The reports on achieved tasks are published to view and evaluate in the group. Furthermore, in the future, we will be able to model the process of development of ideas by machine learning. To that end, what kind of evaluation was used/reconsidered by the idea presented at the meeting, what kind of opinion came out when presented at the meeting again, what kind of evaluation was given when the task derived from the idea was achieved. Based on the results of this analysis, we will construct a machine learning model to estimate the creativity of ideas.

By applying this model to the initial idea, you can predict the future development of the idea. Evaluation of idea creativity can be used to set priorities for multiple ideas for similar problems. I believe that people can accelerate innovation by working according to this priority. I will talk more about this in Chap. 6 of this book.

As we will return to in Chap. 6, our purpose is not just to develop the system to support humans by recording meetings and discovering knowledge. Another goal is to support the general creative activities of humans and to create an environment that enables humans to perform more creative work more efficiently. At the same time, we are thinking about a mechanism that humans can extend their potential and that the system can improve its accuracy simultaneously. Then, the expansion of human capability will be a positive cycle.

In other words, post-meeting assistance contains two main items. One is to make the participants utilize the knowledge gained at the meeting for the subsequent activities, and the other is to evaluate and cultivate discussion ability of the participants through meetings.

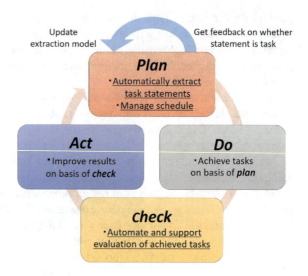

Fig. 3.6 Flow of task achievement

As we discuss the evaluation of discussion ability and its improvement support in Chap. 4 in detail, we would like to talk about the mechanism to make use of the results of the meeting for subsequent actions.

The flow of task achievement is outlined in Fig. 3.6, and the underlined parts are executed by the system. First, the user creates notes for a performance plan to complete the tasks for the task statements extracted by using the learned extraction model and then manages the execution schedule with the scheduling components of the system. He/she then carries out the tasks according to the planned schedule, receives evaluations from other users (support for the check step) on the basis of the results added to the performance plan notes, and considers the evaluations to improve subsequent activities.

As an information tool to use after the meeting, there is an activity note creation application as shown in Fig. 3.7. This is a system to select and review the task statements estimated by the machine learning model trained by the contents structured by the meeting recorder and the DM system, to confirm the tasks, and to create, edit and organize the activity notes. The user can also classify past meetings by topic and select and review some important statements.

Activity notes can include quotation of related task statements and organize items to achieve tasks found during and after the meeting. The task statement is associated with the discussion content, so the user can also view the surrounding statements. The task statements quoted by this tool are registered as correct answer data and are used for relearning. Active learning as described in Chap. 2 allows the user to select a statement to which a teacher signal should be given. By emphasizing the candidates of task statements on the tool, it can prompt the user to browse and quote them. Teacher signals for incorrect answers will be given to statements that were not quoted even after some time.

3.3 Post-meeting Assistance to Support Creative Activities

Fig. 3.7 Activity note creation tool

Fig. 3.8 Task scheduler

When creating an activity note and planning to do items in the note, the user needs to check his/her schedule. Figure 3.8 is a tool called the task scheduler that allows the user to incorporate the contents of activity notes, to correlate them to the appropriate dates and times, and to check other activities and tasks underway. The pie charts on the upper left of Fig. 3.8 show the percentage of time required for the activities of this week and this month. In addition, the vertical bar chart below shows the number of accomplishments per month, and the right bar chart shows the cumulative task achievement status for all tasks. These data can be used not only to plan activities, but also to evaluate activities and to know their own skill level.

Research is a typical example of an intelligent creative activity. In many cases, research will first decide the theme, think about the next goal in that theme and the issues related to that subject. The important thing in solving the problem is that it achieves some goal. In other words, it is necessary to take into consideration which goal is set in detail in advance, which element of that goal is tied to the tasks to be carried out and achieved by that solution.

The state of awareness of students with little experience in research activities who only use the number of tasks is disadvantageous because of the uncertainty regarding execution. Therefore, after the existence of tasks is identified, each task should be well organized and scheduled to use time efficiently. The task scheduler works well for scheduling the duration of task executions.

There is a graph at the top of the screen of the task scheduler that can roughly organize the proportion of information by type (e.g., surveillance, development, and experiment) of tasks and their situation with achievement. The user can arrange and update the time intervals of tasks scheduled to be executed on the timetable at the bottom of the screen.

Also, if multiple activities are to be performed in parallel, it would be better if the planning is done after checking whether there are dependencies on tasks to be performed and considering priorities. For example, if it is assumed that implementation of program X is necessary to carry out task A, and that program X can be implemented by modifying program Y, which is a result of another task B, then task A has a dependency to task B. In that case, it is more efficient to achieve task B earlier than task A. In order to discover such dependency between tasks, we analyze the activity notes and the quoted task statements and the discussion contents, because they are good resources to find clues about task dependency.

As a way to gain insight into dependency, we analyze the similarity of topics, the commonality of specifications or functions of anticipated results, relationships between the original idea and derived discussions, and so on. Anyway, if we continue discussion at meetings, the discussion content should definitely include data that can be used for reasoning of task dependency.

The most important thing in executing a task is that it achieves some goal. In other words, it is necessary to first consider what target is set, which part of the goal is associated with the task to be carried out, and what will be achieved by the solution.

The creative activity support system has a map that roughly describes the entire activity in tree structure form [called mind map (Buzan 1990, Beel and Langer 2011)] and can search which part of the target is related to the problem to be solved from the map as shown in Figs. 3.9 and 3.10 (Fig. 3.10 is an enlarged center of Fig. 3.9).

Then, the user can search on the map which part of the goal is the issue to be solved. By doing so, the users can continue their activities while confirming that they are heading toward the goal by doing so, and not just executing the task in front of them.

We must begin by being aware of the problems that confront us at the moment to facilitate creative activities. As mentioned above, task statements extracted with the proposed system are presented in Fig. 3.11.

3.3 Post-meeting Assistance to Support Creative Activities 69

Fig. 3.9 Activity mind map

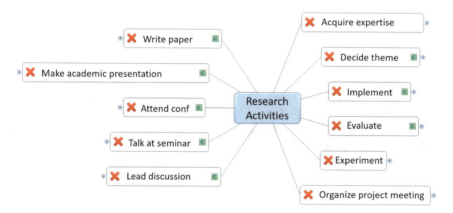

Fig. 3.10 Central part of activity mind map in Fig. 3.9

Each statement that is presented is determined to be a task statement with high levels of possibility, and it is necessary for the user to finally assess whether the task should be achieved. Therefore, since the statement immediately after extraction has a blue icon marked "Task?", which can be clicked, it determines whether the statement indicates a task to be achieved.

By using the activity note creation tool mentioned above, the user creates a note citing the task statement after deciding the task and describes the details of the plan and results related to the task. By setting the attribute of this note to "Finished," a completion icon is displayed in the related task statement. The user can then easily be aware of which task is in a completed state.

Fig. 3.11 Presentation of tasks

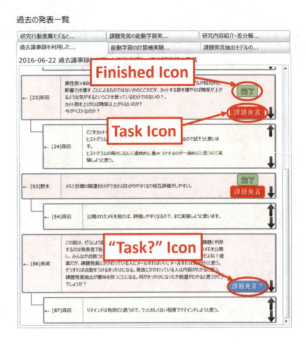

After the tasks are carried out, we should assess their progress, comprehend the level of achievement, and use it as a source for future activities (Nagao et al. 2017). We implemented a function to evaluate task execution content in the proposed system by obtaining other students' assessments of the content of execution by publishing task execution notes in the lab.

We confirmed that mutual evaluation was reasonable by referring to the activity notes that described the content of task execution. However, since the task execution notes were primarily private, they are not supposed to be published. The sharing function of the notes provided an opportunity to receive a more accurate evaluation when there was motivation to tell others about the results of their task execution.

When completing task execution, the system accesses the level of achievement of the research goals. The user can then publish related research notes and associate them with the previously shared notes to motivate others to update evaluations. The associated notes are displayed on the right of the evaluation window, as seen in Fig. 3.12, when receiving an evaluation from another user. The user considers them as a supplemental resource to evaluate another's research results.

In the creative activity support system, there is also a function that can receive evaluation according to the content by publishing the activity notes as a task execution report in the group as the evaluation function of the user's activity concerning task execution. The screen example shown in Fig. 3.12 is an interface for that, by clicking on one of the five stars, it evaluates the degree of achievement and evaluates the effort of the user (even if the result is not good to evaluate), the user can click on the hand icon (this is called stamp evaluation).

3.3 Post-meeting Assistance to Support Creative Activities

Fig. 3.12 Task evaluation tool

The notes describing the activity contents are generally private, and do not assume that they are made public. The note publication function is to provide opportunities to receive evaluation from others when there is an intention to tell others about the result of their task execution. This is a mechanism that adopts the gamification described in the next chapter.

In addition, the creative activity support system has a function to automatically evaluate the activity process based on task statements, activity notes, and related goals. For this, several numerical targets are set in the abovementioned map, and the note has a column for describing the progress status corresponding to those numerical targets. A numerical target is like, for example, investigating more than five related instances of research. At this time, it is possible to notice the user by displaying at all times what percentage of the target is achieved. Also, by calculating the degree of relevance between the task statements and the content of the activity notes, it is possible to evaluate the degree of relevance to the contents of the activity with the contents of the challenge that triggered it. Tasks and activities are linked correctly, and furthermore, by using numerical targets as clues, we will be able to progress creative activities more steadily.

By confirming the goal based on the map that bird's-eyed the activity and continuing activities while utilizing the PDCA cycle, you will be able to achieve results steadily. However, creation activities that are extremely difficult to create maps for are likely to exist. As aiming at the development and practical application of IT products that make commoditization (to become necessities for daily life) next to smartphones, it will be difficult to clarify what kind of map to work on. For research activities, there is some standard way to proceed, but if you try to make innovative inventions, it may be difficult in the same way as other people. Sometimes there is a doubt that activities that can make maps are not so creative in the first place. However,

Fig. 3.13 Top page of creative activity support system

what we can say from our experience is that maps and guidelines (procedures of what to do) are effective in doing research, and that you can demonstrate creativity as well.

In our laboratory, the members logged into the creative activity support system once a day, confirmed the activities of the day, checked the information published by others, they did it without forgetting they are trying to describe things. Figure 3.13 shows the top page of the system after login. From this page, the user can check their own and others' tasks and their situation and view the details. It is also linked to the discussion browser mentioned in Chap. 2 and the personal page of gamified discussion explained in Chap. 4 (detailed confirmation page of students' discussion ability), and the state of themselves at the meeting. So, the students can learn more about the current situation with it.

We are planning to develop a system to discover creative ideas that lead to future innovation by efficiently recording and analyzing meetings and subsequent activities, and to develop it efficiently. Future plans also include clarification of the true value of this research by gathering data on a large scale, building a machine learning model and operating it.

We predict that most of the intellectual creative activities will be carried out by the group, and it will be in the form of activities being promoted during discussion. In other words, it is the idea that support of group activities centering on discussion will directly support human creative activities.

As it is no longer an age that some genius thinks of everything, it will be that people will share wisdom, invent and discover. At that time, what is important is a system for recording and effectively utilizing discussions and results.

In this way, in order for an artificial intelligence system that combines machine learning and natural language processing to support human creative activities, it is necessary for human beings not to hide themselves and to enter information about their activities into the system.

3.4 Evaluation of Creativity

Although how to capture creativity is varied, we regarded it as the degree of development of ideas in this research. In other words, the extent to which ideas were refined and materialized after discussion was used as an indicator of creativity.

The ideas should appear in the statements in the discussions and be extracted from the minutes and cited in the creative activity support system. Their further development will therefore be possible. Then, after the reports are published and the evaluations are received, the values of the ideas are augmented, and they can be further developed by making them the topics of the next meeting.

Creativity can be represented by using concrete data and modeled by machine learning in this way by tracing the generation of ideas through to their development.

The idea seems to have something of value at the time of its creation, but it is very difficult to evaluate, so by presenting it at a meeting and placing it on the top of the discussion, the idea of later development will leave the possibility and direction to the judgment of participants. Of course, depending on the experience and knowledge of the participants, it is also good that the idea creator compares reactions by showing similar ideas at meetings attended by other participants.

In this way, by tracing the development process of ideas, it is possible to represent creativity with concrete data, and it can be modeled by machine learning.

The development of ideas is not necessarily carried out by a specific individual. Currently, only the activities of individual users registered in the creative activity support system can analyze the development process of the idea in detail, and only by discussion, multiple users can develop ideas (whether idea should be developed or not can involve judgment). However, there are cases where multiple people will integrate and develop ideas that come up separately. It is necessary to analyze collaborative idea development of multiple users in the same way.

When it becomes possible to predict the creativity of ideas by machine learning, and when multiple ideas to solve the same problem are conceived, it becomes possible to rationally decide which one to develop preferentially. This is done by presenting multiple ideas at a meeting and by automatically evaluating extracted statements after discussions.

It is thought that innovation will be accelerated by repeating choices based on the creativity of the ideas. Innovation cannot be achieved unless it develops ideas to a level that can affect society, but it will be possible to efficiently use time to carry out creative activities by considering which ideas to preferentially develop.

This is presently not an era when innovation is generated from a few outstanding ideas. However, a plethora of ideas that can create innovation in the future can be created from discussions by numerous people. True innovation can be achieved by carefully developing such a profusion of creative ideas.

It might be the case that humans might become more creative through the support of artificial intelligence. Whether or not creativity itself can be entrusted to artificial intelligence is a matter that is likely up for debate, but artificial intelligence can be used in places where humans are not very good and will help humans in various ways, for example, to discover high-priority items from a large amount of information and many others.

The thing that humans should do is to record the discussion firmly and make analysis possible. After that, we will conduct creative activities based on goals, issues and plans, and keep track of the details as much as possible. I think that recording of activity is very troublesome unless the person taking the record is in the habit of doing it (such as a person writing a diary every day), but if they are trying to make their activities more meaningful and creative, it would be better to adapt to that way.

3.5 Future of Creative Meeting

Since a meeting is an important means of communication and a place for multiple people to share ideas, it is natural to try to acquire various knowledge using the minutes, which is the record, as a knowledge source. The more knowledge the participants can earn, the higher value of their meeting can produce. One way to use the minutes for multiple purposes is a mechanism that creates answers to any question of the user by combining information obtained from the minutes. This is called an interactive minute. In order to realize the interactive minutes, detailed semantic structuring of the minutes is necessary, and it is expected to realize advanced artificial intelligence technology (e.g., question answering system).

In order to facilitate collaborative work, a mechanism called a shared editor has been proposed based on the idea that all the participants need to be able to manage any information which everyone is seeing in the shared space. In this mechanism, since multiple participants write and view the information at the same time, exclusive control and so on need to be devised. Therefore, it is possible to set a writable layer for each participant, make it overlap, and at the same time use the color-coded pointers to grasp the points indicated by each participant.

In the future, pseudo-face-to-face meetings using virtual space will be heavily used. In the meeting, participants are held by wearing a virtual reality headset and virtually gathering in a place simulating real meeting rooms or halls. There, various materials will be presented using video/audio/3D data, and discussion will be carried out by voice accompanied by text subtitles. Of course, the minutes will be automatically created and their summaries will be generated.

An anthropomorphic agent as a facilitator emerges in the virtual space, in some cases proposing to prevent divergence of the discussion. It also proposes to make an appropriate break during the meeting, and extracts keywords that seem to be important. The agent then arranges the keywords in order and shows their diagrammatical representation. It also supports comparisons with past meetings and supports or disapproves the current proposal based on the comparison results.

However, the main participants of the discussion will still be humans. Human creativity cannot be imitated sufficiently by artificial intelligence yet. This is because the current artificial intelligence is a mechanism to learn from the data. Humans do not necessarily think based on data. For that reason, there are things that could lead to a mistake, but on the contrary, there is a chance to make a revolutionary idea that cannot be deduced from existing data.

By making a meeting in the virtual space, physical contact becomes difficult, but the advantage of not having to travel long distances is very great. Moreover, almost the same situation as the face-to-face meeting is reproduced, so the nuance of words is easily transmitted. Although it is difficult to recognize the face wearing the headset, appropriate (sometimes deformed) facial expressions are expressed by avatars trained by machine learning in advance.

At such a meeting, the technology we developed will maximize its function.

The techniques of analyzing and structuring meeting and discovering knowledge by discussion mining lead the essence of discussion and function for users to look back past discussions efficiently. In addition, it is possible to reproduce the discussion as faithfully as possible with the technology that automatically collects and uses various data concerning meetings such as a meeting recorder. The contents of past creative activities are not lost, they are reused appropriately. The gamified discussion method explained in Chap. 4 works not only to improve motivation for discussion but also to promote qualitative evaluation of the user's discussion ability. The mechanism for improving the ability of the discussion will also be described in detail in Chap. 4.

References

J. Beel, S. Langer, An Exploratory Analysis of Mind Maps, in *Proceedings of the 11th ACM Symposium on Document Engineering*, pp. 81–84 (2011)

T. Buzan, *Use Both Sides of Your Brain* (Plume Books, 1990)

K. Nagao, Meeting Analytics: Creative Activity Support Based on Knowledge Discovery from Discussions, in *Proceedings of the 51st Hawaii International Conference on System Sciences (HICSS 2018)* (2018)

K. Nagao, N. Morita, S. Ohira, Evidence-based education: case study of educational data acquisition and reuse. J. Syst. Cybern. Inform.: JSCI. **15**(7), 77–84, ISSN: 1690–4524 (Online) (2017)

T. Osone, K. Uota, An Approach to Teaching Basic Information Education based on PDCA Cycle. Bus. Rev. Senshu Univ. **100**, 1–14 (2015)

Z. Qu, Y. Liu, Sentence Dependency Tagging in Online Question Answering Forums, in *Proceedings of the 50th Annual Meeting of the Association for Computational Linguistics*, pp. 554–562 (2012)

Chapter 4
Discussion Skills Evaluation and Training

Abstract There must be as many concrete indicators as possible in education, which will become signposts. People will not be confident about their learning and will become confused with tenuous instruction. It is necessary to clarify what they can do and what kinds of abilities they can improve. This paper describes a case of evidence-based education that acquires educational data from students' study activities and not only uses the data to enable instructors to check the students' levels of understanding but also improve their levels of performance. Whether a meeting is executed smoothly and effectively depends on the discussion ability of the participants. Evaluating participants' statements in a meeting and giving them feedback can effectively help them improve their discussion skills. We developed a system for improving the discussion skills of participants in a meeting by automatically evaluating statements in the meeting and effectively feeding back the results of the evaluation to them. To evaluate the skills automatically, the system uses both acoustic features and linguistic features of statements. It evaluates the way a person speaks, such as their "voice size," on the basis of the acoustic features, and it also evaluates the contents of a statement, such as the "consistency of context," on the basis of linguistic features. These features can be obtained from meeting minutes. Since it is difficult to evaluate the semantic contents of statements such as the "consistency of context," we built a machine learning model that uses the features of minutes such as speaker attributes and the relationship of statements. We implemented the discussion evaluation system and used it in seminars in our laboratory. We also confirmed that the system is effective for improving the discussion skills of meeting participants. Furthermore, with regard to skills that are difficult to evaluate automatically, we adopted a mechanism that enables participants to mutually evaluate each other by applying a gamification method. In this chapter, I will also describe the mechanism in detail.

Keywords Discussion ability · Speaking ability · Listening ability · Evaluation of discussion skills · Training of discussion skills · Gamification

---

© Springer Nature Singapore Pte Ltd. 2019
K. Nagao, *Artificial Intelligence Accelerates Human Learning*,
https://doi.org/10.1007/978-981-13-6175-3_4

4.1 Evaluation of Speaking Ability

Analyzing discussion scientifically provides an evaluation of the appropriateness of the statements and the discussion ability of the speaker. Evaluating the discussion ability of the speaker leads to evaluating the person's communication ability. As I mentioned at the beginning of this book, the ability to conduct discussions can be said to be the essence of communication skills, so having a high discussion capability necessarily means a high communication skill.

To evaluate discussion ability, we first evaluate individual statements and evaluate the speaker's abilities based on the evaluation of his/her statements. Our first step is to evaluate an individual's speaking ability.

As explained in Chap. 3, the meeting support system we developed records the statements of each participant during the meeting as the discussion content including video/audio data and text minutes. Therefore, we can automatically evaluate the statements based on their acoustic features and linguistic features.

At the meeting, participants need to discuss a topic, analyze the story of the other persons' statements, and communicate their argument in an easy-to-understand manner. "Pronunciation," "speech speed," "pause," "conciseness," etc. are mentioned as easy-to-understand way of speaking (Kurihara et al. 2007). Based on this, we set eight evaluation indicators. The evaluation indicators are based on acoustic features and those based on linguistic features.

The indicators that evaluate only by acoustic features are as follows:

A. Voice size: voice should be large enough for the speaker to hear enough, while on the other hand it is better not to be emotional and too loud. Therefore, we measure and evaluate the volume [dB] of each statement.
B. Voice intonation: speech without intonation is a factor that makes a listener bored. We measure the height of the voice in the statement (fundamental frequency F0 described later) [Hz], and evaluate those with high standard deviation values as good evaluations.
C. Speed of talking: it will be hard to hear even if the statement is too fast or too slow. Therefore, if the speech speed (the number of syllables per hour, the syllable is described later) is within the appropriate range, it is evaluated as good.
D. Fluency: speech with a lot of silence and disfluency is difficult to understand. A good evaluation is given to statements with few filled pauses (vowel extension) such as "eh" during speaking and few silence periods of more than 2 seconds.
E. Tempo: it seems easy to understand the speech when the emphasized part is clear. It is effective that the statements are not monotone such as speaking slowly the part that you want to emphasize and setting a pause before the emphasized part. Therefore, the tempo of the statement is evaluated based on the standard deviation of the speech speed and the number of "pauses" ("pause" is defined as a silence period of less than 2 s).

Here, the fundamental frequency (generally written as F0) is a value expressing the periodicity of sound, which is the acoustic feature quantity that governs the pitch

4.1 Evaluation of Speaking Ability

of sound. There is periodicity in voiced sound (vibrating the vocal cord), so the reciprocal of that period (basic period) is the fundamental frequency.

F0 is a very important index to consider for the intonation of a voice, but (1) a speech waveform is quasiperiodic signal (due to quasiperiodicity of vocal fold vibration), periodicity is not clear, (2) speech is mixed with noise, and (3) the range of change of F0 in voiced sounds is difficult to limit because of the wide range. Accurate extraction of F0 is very difficult. Therefore, several estimation methods have been proposed. The acoustic analysis program called speech signal processing toolkit (SPTK) has been released (http://sp-tk.sourceforge.net/), in which an algorithm called pitch extraction is implemented to estimate F0.

In addition, the syllable used to calculate the speech speed is a type of segmental unit that separates consecutive voices, and is a group of sounds heard. Typically, it is a voice (group of voices) consisting of one vowel and its vowel alone or with one or more consonants before and after the vowel. In the case of Japanese, syllables may use a segment unit called a mora (beat) that does not necessarily match. Strictly speaking, the mora is used instead of a syllable. The main difference between syllable and mora is that the long vowel, the geminate consonant, and the syllabic nasal are integrated with the preceding vowel in the case of the syllable, but in the case of the mora it is 1 mora.

Next, evaluation indicators based on linguistic features are as follows:

F. Conciseness: it is easier to understand if the statement is concise. Therefore, for the sake of conciseness evaluation, we compare the number of syllables of statements (strictly mora) in the meeting by speech recognition and the number of syllables of the corresponding statements in the minute of the meeting. Since the secretary describes the content of the statements in summary, if the number of syllables of the statements and the number of syllables of the corresponding statements of the minute are close, it can be considered that the statements can be regarded as being concise.

G. Relevance with start-up statements: statements should be relevant with the subject of discussion as much as possible. If the content of follow-up statements is in common with the content of the topic raising statement (i.e., start-up statement), it can be considered that it is relevant with the theme. Therefore, by evaluating the degree of relevance to the start-up statements (described later), the relevance of statements is evaluated.

H. Consistency with parent statements: follow-up statements need to be united or consistent with their parent statements. In other words, the content of the follow-up statement and the content of its parent statement must be semantically related, so it is important to evaluate the degree of the consistency. We use a machine learning technique to decide whether the statement is a "consistent statement or not" which determines the evaluation. The method is described later.

We calculate the degree of relevance between statements in the following way. First, we calculate term frequency–inverse document frequency (TF-IDF) values of words in each statement by using to following formula:

$$tfidf_{i,j} = tf_{i,j} \cdot idf_i$$

$$tf_{i,j} = \frac{n_{i,j}}{\sum_{k \in T} n_{k,j}}$$

$$idf_i = \log \frac{|D|}{|\{d : d \ni t_i\}|}$$

Here, $n_{i,j}$ is the number of occurrences of word t_i in document d_j.

$\sum_{k \in T} n_{k,j}$ is a summation of the number of occurrences of all words in document d_j.

$|D|$ is the number of all documents and $|\{d : d \ni t_i\}|$ is the number of documents that contain the word t_i.

IDF works as a kind of general language filter. If words (generic words) appear in many documents, their IDF values decrease. If words appear only in specific documents, their IDF values raise.

Using the TF-IDF value with each statement of one meeting as one document, we weight the word t with the following formula to obtain the degree of relevance:

$$f(t, d_1, d_2) = \left(\frac{tf(t, d_1)}{\sum_{s \in d_1} tf(s, d_1)} + \frac{tf(t, d_2)}{\sum_{s \in d_2} tf(s, d_2)} \right) * idf(t)$$

For words that appear commonly in two documents, add that value, subtract the value for words that appear only in one side, and sum over all the words. Let this value be the degree of relevance between statements. An example of the calculation is shown in Fig. 4.1.

In Fig. 4.1, $t_1, \ldots t_6$ are words and d_1, d_2 are statements. In this example, t_1 and t_2 appear in common, other words appear only in one. For common words, the weighted values by the TF-IDF value are summed to 6.0, otherwise the total weighted value by the TF-IDF value is added to 2.5, so the relevance is $6.0 - 2.5 = 3.5$.

By the way, it can be said that inconsistent statements in the discussion are statements that describe topics that are different from topics up to that point. So, consider how to categorize follow-up statements as statements deviating from topics or not. Logistic regression analysis described in Chap. 2 is used for this classification. In this case, calculate the probability value that the statement is deviated from the topic, and use this value for the consistency evaluation of the statement. For this purpose, in addition to the linguistic features obtained from the minutes, we use the meta-information given to the minutes. The features used in this method are as follows:

(1) Features based on linguistic features of statements.

- Relevance to parent statement.
- Number of sentences of statements.
- Number of characters of statements.
- Morpheme unigram and morpheme bigram (see Chap. 2).
- Presence of subject word and referring word.
- Entity Grid (see Chap. 2).

4.1 Evaluation of Speaking Ability

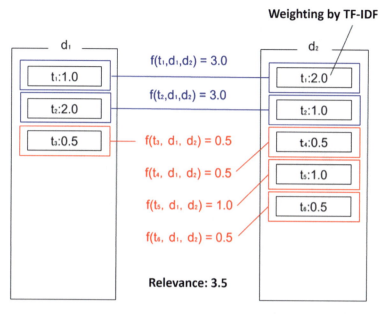

Fig. 4.1 Relevance between statements

(2) Features based on meta-information attached to the minutes

- Whether the speaker is a student or not, whether it is a presenter or not.
- Whether the speaker of the parent statement is the presenter.
- Presence of marking/agreement/disagreement buttons.
- Depth from the root in the tree structure (i.e., discussion chunk).
- Whether or not the visual referent of the parent statement matches that of the target statement.
- Presence or absence of slide operation during speaking.
- Time for reservation of speaking.
- Presence or absence of different statements in time series between the parent statement and the target statement.
- Alternation of questioner.

For morphemes and morpheme pairs that appear during speech, the number of occurrences of nouns, verbs, adjectives, auxiliary verbs, and morpheme pairs is calculated by preliminary survey as in the analysis of the task statements described in Chap. 2, we used those exceeding a certain value for the feature. Also, since there is a report that Entity Grid mentioned in Chap. 2 is effective for evaluating text consistency (Barzilay and Lapata 2008), it is directly related to topic transition among the syntactic role of the Entity Grid. We focused on only the transition of the theme considered as a transition probability and used it for the feature.

	Precision	Recall	F-measure
Proposed method	0.269	0.534	0.358
Comparative method	0.117	0.129	0.123

Fig. 4.2 Experimental results

The alternation of the questioner, which is the last feature where we separate into questions groups by whether or not the same person asks two questions in a row to the presenter, or whether there are two different people in a row that ask the presenter a question.

We implemented the above method and conducted an experiment on discrimination of inconsistent statements. As a dataset, we used 53 min (discussion content) of seminar in our laboratory (number of statements: 3553). However, since the start-up statements are not the subject in this case, the follow-up statements (the number of statements: 2490) are subject to discrimination. As correct answer data (teacher signals), we decided manually whether a certain statement lacked consistency, and gave the attribute of inconsistency to the statement. 202 follow-up statements were determined to be lacking in consistency.

In order to evaluate the proposed method, the case where learning was carried out without using features based on the meta-information of the minutes was taken as a comparative method. For the evaluation, we used the precision, the recall, and the F-measure that is a harmonic mean of these two values, and additionally carried out the tenfold cross-validation.

The results of this experiment are shown in Fig. 4.2. The results of the consistency judgment by the method we proposed are higher than the case where the feature information given to the minutes is not used, for all the precision, the recall, and the F-measure. The advantage was confirmed.

In addition, when learning by removing each feature by the meta-information of the minutes, the precision, the recall, and the F-measure declined in all the features, and the effectiveness of the used feature was confirmed. Figure 4.3 shows the results of the top five cases where the F-measure drops greatly.

We will automatically evaluate all statements of the meeting participants using the evaluation indicators mentioned above. Then, let the weighted average value of the value of each indicator be the evaluation of one statement, and let the sum of the evaluation values of all statements of the participant be the numerical value expressing that participant's speaking ability in discussions at that meeting. By looking at the changes for each discussion at each meeting, the participants will be able to judge whether their discussion skills are rising or stagnating.

4.2 Feedback of Evaluation Results

Removed feature	Precision	Recall	F-measure
Whether speaker is presenter	0.255	0.494	0.337
Presence of marking/agreement/disagreement buttons	0.251	0.522	0.341
Depth from root in tree structure (i.e., discussion chunk)	0.253	0.522	0.341
Whether visual referent of parent statement matches that of target statement	0.259	0.506	0.342
Alternating of questioners	0.255	0.534	0.345

Fig. 4.3 Effectiveness of features

4.2 Feedback of Evaluation Results

The evaluation indicators as described in the previous section are indices for measuring the participants' speaking ability, but of course it should be used not only to measure but also to extend their ability. One of the way to do this is to visualize the results in an easy-to-understand manner and feed back to the participants at just the right time.

The participants should make efforts to raise their speaking ability. For that purpose, the system we developed evaluates their statements during the meeting, points out the problems, and encourages improvement. There are various ways this is pointed out. One is to display a message on the main screen during or shortly after speaking or to display the icons next to the name of each participant in the member table of the subscreen. There is another way to display feedback including somewhat detailed information like the icons and their descriptions shown on the tablet used by all the participants. Let me explain each.

The evaluation indicators through using the acoustic features described in the previous section can automatically calculate the evaluation values and feedback during the meeting. Specifically, they evaluate in real time each of "voice size," "voice intonation," and "speech speed", and when it is a value lower than a certain threshold value, that is, a "bad" evaluation value, the system pops up a warning message immediately on the main screen (normally displaying the presentation slide) as shown at the bottom right of Fig. 4.4. This display will be hidden after 2 s.

In order to measure the effect of this simple direct feedback on participants, we evaluated the participants' "voice size," "voice intonation," and "speech speed" at five meetings. Results of examining the change in the evaluation values are shown in Fig. 4.5. For "voice size," the message to be displayed differs depending on whether the evaluation value is smaller or smaller than the reference value. Regarding "speech speed," it may be faster or slower than the reference value, but in the preliminary experiments, it was extremely rare when it was later than the reference value, and it

Fig. 4.4 Feedback message pop-up on main screen

Evaluation indicator	Message	# of displays	# of improvements	Improvement rate
Voice size (small)	Speak loudly	32	27	0.84
Voice size (large)	Calm down	5	4	0.80
Voice intonation	Speak with inflection	30	20	0.67
Speech speed	Speak slowly	67	44	0.66

Fig. 4.5 Experimental results of effects of feedback on main screen

was overwhelmingly more in the case of the faster case. Only when the evaluation value is larger than the reference value, the message is displayed.

Improvement numbers shown in Fig. 4.5 are counted as the total number of improvements for all participants. Improvement means that the evaluation value concerning the indicator of a certain participant becomes better after a message on the indicator is displayed while the same person is speaking or immediately after speaking. The improvement rate is the number of improvements for a certain indicator divided by the total number of messages in the indicator. In the five meetings, we could not collect sufficient data, but we found trends of improvement by feedbacking the evaluation results of the statements based on the acoustic features in real time.

In addition to the main screen, there are two subscreens for displaying meeting metadata in discussion mining so that participants can view one of them at any time during the meeting. In addition to the statement content (summary) described by the secretary, the tree structure of the discussion currently being created and the reservation status of the statements are displayed.

4.2 Feedback of Evaluation Results

Fig. 4.6 Feedback display on subscreen

Then, as shown in the part surrounded by the red frame line in Fig. 4.6, we present the evaluation result on the latest statement with the icons next to the participant's name on the panel on which the meeting participant list is displayed. The icon type corresponds to each evaluation indicator (voice size, speech speed, voice intonation, fluency, and tempo), the icon color corresponds to the result (good: green, normal: yellow, and bad: red). The icons to be displayed are limited to two in terms of space and visibility. One of the combinations that are two poor (or ordinary) evaluation indicators or one good evaluation indicator and poor (or ordinary) evaluation indicator is displayed.

Unlike the main screen, the subscreen does not always come into view, so it seems that there is not a direct effect. However, the feedback display to the main screen disappears immediately, whereas the display on the subscreen is kept until the next statement by the same participant, it is considered to be effective if the participants want to know their evaluation results compared with the others.

In the discussion mining (DM) system and the meeting recorder, each participant in the meeting is using the tablet. In the DM system, the tablet displays the same thing as the presentation slide displayed on the main screen, and the user can mark it with a pen or finger.

Fig. 4.7 Feedback display on tablet

On the tablet, as shown in Fig. 4.7, using the five icons, the result of each evaluation indicator for each statement is displayed immediately after the end of the statement. As with the above-described subscreen, the type of icon corresponds to each evaluation indicator (voice size, speech speed, voice intonation, fluency, and tempo), the color of the icon corresponds to the result value. The evaluation indicators based on the linguistic features are still difficult to analyze in real time, so they are not shown on the tablet.

Since the display on the main screen and the subscreen is carried out in advance of the display on the tablet, the independent effect on display on the tablet is not measured. However, in order to improve fluency and tempo, we found that it is necessary for the participants to practice their speaking style.

As mentioned in Chap. 2, discussion mining uses a system called a discussion browser for viewing discussion content. Also, in the meeting recorder of Chap. 3, there is a viewing system for the minutes as well. Even on these systems, the user can check the content of each statement and the evaluation result of the statement.

As shown in Fig. 4.8, eight icons are displayed together with other metadata directly under the part where the statement content is displayed. At this time, the content is also analyzed, and the results are also displayed for the evaluation indicators based on linguistic features. Three icons at the right end of the icon row correspond to them. From the left, it shows the result of simplicity (square icon), relevance with start-up statement (icon of "subject"), and consistency with parent statement (icon of "flow").

Icons are superior in quick overview, but are not appropriate to know the details, so if the user moves the mouse cursor over the icon, a description pop-up will be displayed (such a user interface is generally called a tooltip). An example is shown in Fig. 4.9. In this example, if the user moves the cursor to the icon in the form of a wave, the description "voice intonation" and the advice "let's give a little more inflection" are displayed.

Although it seems that there is immediate effect in the feedback of the evaluation results during the meeting, it may be difficult for the participants to continue speaking at the next meeting while remembering their weaknesses pointed out last time. That's because the participants are not always trying to raise their speaking ability which

4.2 Feedback of Evaluation Results

Fig. 4.8 Feedback display on discussion browser

Fig. 4.9 Tooltip for description of evaluation

eventually improves discussion skills, and they must pay attention to other issues to be considered among meetings (e.g., achievement of tasks).

For this reason, a mechanism to remind the user of the problems of the statement at the last meeting is required. Of course, if the user reviews the minutes, he or she can reconfirm the evaluation results of the statements as well as the contents of the previous meeting, but it is unlikely that he/she will frequently review the minutes unless it is a very important agenda.

In the past, we implemented a mechanism of mail notification to let participants know that the minutes were completed and accessible. Apart from that, this time, we added a mechanism to notify the participants by the result of the evaluation on the statements at the last meeting and the points to be noticed in the next meeting by e-mail.

An example of feedback by this mail is shown in Fig. 4.10. This is called HTML mail, and the receiver can display contents including images and links to Web pages on the mail application. The sentences and graphs shown in Fig. 4.10 were automatically generated based on discussion data. Compared with the evaluation results of the previous meetings, the mail comments on the items that show little improvement with referring the data.

The timing to send this mail is just after the minutes are ready to browse and around noon of the day when the meeting is held the next day. We have a meeting using the DM system every Wednesday, so we will receive two e-mails within a week.

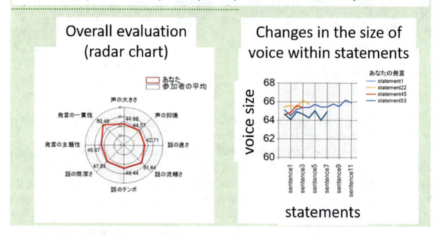

Fig. 4.10 Example of feedback mail

4.3 Evaluation of Listening Ability

Of course, only the speaking ability that I mentioned in the previous section is not the ability of discussion. Since discussion ability is a complex ability, it can be analyzed from various viewpoints.

Well, while considering "speaking ability," we will need to consider "listening ability" as well. There are also indicators that do not raise the evaluation value unless the participants cautiously listen to the statements of other participants such as relevance with start-up statements and consistency with parent statements. However, for evaluating from the viewpoint of understanding the entire discussion, it is considered that these indicators are insufficient.

Therefore, we propose "summarization ability" as ability equivalent to "listening ability." This is the ability to hear people's statements and summarize them. In the DM

4.3 Evaluation of Listening Ability

system, students who are in charge of secretaries during the meeting write content of all statements. However, since it is very difficult to enter all the contents of the statements, it is inevitably the way that the secretary enters while summarizing the statements. We assume that summarization ability is an approximation of the ability of understanding discussions.

The evaluation indicator of the summarization ability is not mere description quantity. Even if a person with a fast keyboard input can enter many characters in a short time, it does not mean that their summarization ability is high. Besides that, it is said that a person who enters only the contents considered necessary and does not input contents judged to be redundant is more summarized.

So, we conducted an experiment to check the following items on the text of the statements entered by the secretary (Tsuchida et al. 2010).

(1) Individual differences in description quantities by secretary.
(2) Relationship between the type of discussion and the description quantity.

In this experiment, we used the discussion content which the six undergraduate and graduate students of our laboratory created for the discussion content, we compared the statement text (hereinafter referred to as the secretary text) entered by the secretary and the full spoken text (hereinafter referred to as transcription text) which was generated by using the speech recognition of the meeting recorder and then corrected for recognition errors by humans in terms of the proportion of keywords.

To take account of individual differences in the written amount of words by the secretary, we sorted six secretaries into three groups (two each for each) according to the average number of input characters per statement. Groups A, B, and C are grouped from the order of the average number of input characters. And, we chose two contents for each secretary in charge. In other words, we targeted 2 discussion contents for each secretary, 4 for each group, and 12 discussion contents overall.

The total number of statements included in all discussion contents was 661 statements (55.1 statements per meeting), and the total number of discussion chunks (discussion trees with start-up statements as the root) was 137 (11.4 per meeting).

The speech recognition results of the discussion content include speech disfluency such as "um" and "ah," but it is deleted in the transcription text. Next, in order to obtain the proportion of keywords in statement texts, morphological analysis was performed on each secretary text and transcription text. The type and number of parts of speech included in the result are shown in Fig. 4.11.

The TF-IDF value as described above was calculated using a self-contained word (a word making a meaning by itself) obtained by morphological analysis as a keyword candidate, and one having a threshold value or more was taken as a keyword of a statement.

The total number of characters in the discussion content used in this experiment was 41,946 characters (3495 characters per meeting) in the secretary text and 188,816 characters in the transcription text (15,734 characters per meeting). Also, the number of morphemes (including duplicates) obtained by morphological analysis was 23,950 in the secretary text (1995 per meeting) and 105,843 in the transcription text (8820 per meeting). From the viewpoint of the number of characters and the number of

POS	Secretary text	Transcription text
Noun	2960	5791
Verb	845	1929
Unknown	359	679
Adjective	169	349
Adverb	24	49
Adnominal	16	49
Interjection	0	1
Other	0	0
Total	4373	8847

Fig. 4.11 Type and number of parts of speech in statement texts

morphemes, we found that the amount of descriptions necessary to transcribe all statement contents is about four times the descriptive quantity entered by the secretary in real time.

For each group of secretaries, the number of independent words per statement in the secretary text (1), the number of keywords per statement in the secretary text (2), the number of independent words per statement in the transcription text (3), the number of keywords per statement in the transcription text (4), and the total number of common keywords (5). The relationship between these data is as shown in Fig. 4.12.

For the following explanation, (2)/(4) shows the ratio of the number of keywords in the secretary text and the number of keywords in the transcription text which is used as an indicator for measuring how much the keywords transcribed the statements. We call it the "transcription rate" and also treat (5)/(4) which shows the content ratio of common keywords for keywords in the transcription text as an indicator to measure how accurately the secretary grasped the content of the statements, we will call it the "grasping rate." Likewise, we call (2)/(1) which indicates the content rate of independent words for the keywords in the secretary text as an indicator to measure how much the secretary has described redundant content. We call it the "reduction rate."

In the experiment, the average of the grasping rate was 43.7%. From this, we can see that more than half of the keywords of the speech contents are missing when the secretary enters them.

4.3 Evaluation of Listening Ability

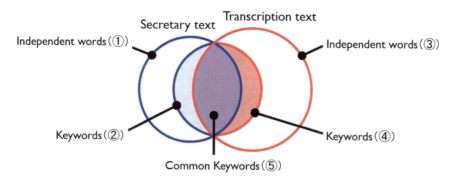

Fig. 4.12 Relationship between independent words, keywords, common keywords in secretary text and transcription text

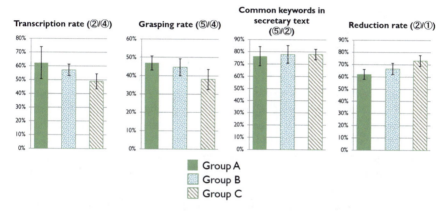

Fig. 4.13 Difference in description amount of statement text for each secretary

Figure 4.13 compares the average and standard deviation of the transcription rate (2)/(4) for each secretary group. Similarly, the content ratio of the common keywords in the transcription text (grasping ratio) (5)/(4), the content ratio of the common keywords in the secretary text (5)/(2), and the content rate of the keywords in the secretary text (reduction rate) (2)/(1) are also shown in the same figure.

Since we grouped according to the average number of input characters per statement, we found that there is a difference in the transcription rate for each group. In addition, from the graph of the grasping rate, it is understood that as the transcription rate increases, more keywords in the speech contents are described. On the other hand, when looking at the graph showing the ratio of common keywords in the secretary text, there was not so much individual difference. In other words, regardless of the amount of text entered by the secretary, the percentage of keywords in the secretary text can be thought of as nearly constant.

From this, it seems that there was not much individual difference in the quality of the minutes as it is, although the amount of text to be entered depends on individual

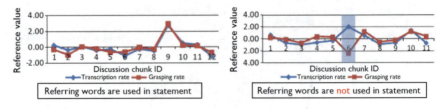

Fig. 4.14 Relationships between transcription rate and grasping rate

secretaries. Also, from the graph of the reduction rate (2)/(1), it can be seen that there is a possibility that the secretary with few average input characters may have entered the secretary text more concisely compared to the secretary that has many. In other words, while a secretary with many average input characters was trying to describe the details of the speaker's speech as much as possible, the secretary with a small average number of input characters was thought to have entered input while paraphrasing the speech content as short as possible.

At the meeting, there are various types such as discussion with only question and answer, discussion such as argument on a certain subject like brainstorming. Therefore, for example, if a meeting has a seminar style that includes a question and answer session, a questioner can refer to a presentation slide, so it is easier to understand the meaning and it is easy to create a secretary text. On the other hand, a discussion including conceptual opinions which cannot be referenced easily through materials can be difficult to understand, which makes it difficult to create a secretary text. It seems that there is a relation between the discussion type and the description quantity of the secretary text.

For this reason, a transcription ratio and a grasping rate are calculated for discussion chunks within each discussion content, a reference value SS_i expressed by the following equation is calculated for each value, and these tendencies are compared. Where X_i is the value of X (X is the transcription rate or the grasping rate), $E[X]$ is the arithmetic average of X, and SD[X] is the standard deviation of X.

$$SS_i = \frac{X_i - E[X]}{\mathrm{SD}[X]}$$

The result is shown in Fig. 4.14. As shown in this figure, in most discussion chunks, the transcription rate and the grasping rate show similar tendencies.

However, as in the sixth discussion chunk in the graph on the right side of Fig. 4.14, there was the case where the grasping rate was small despite the large transcription rate. When we examine this discussion chunk in detail, we found out that it is a discussion about the figure in the presentation slide. In such a case, since the participants are pointing at the figure while speaking, there was a tendency to frequently use referring words. Therefore, while the secretary often inputs a word that supplements the contents of the referring words, it was found that the grasping rate decreases because the words are not directly included in the speech content.

4.3 Evaluation of Listening Ability 93

As mentioned at the end of Chap. 2, the discussion mining system allows the user to select arbitrary character strings in the slide using a device called a discussion commander, so that characters in slides are automatically inserted into the editing window of the secretary tool. This solves the problem of the reference to the slide, so there is no big difference in the description amount of the secretary by the type of discussion.

As there is no dependence on the discussion type and the amount of description of the secretary, the summarization ability can be defined as generically as possible regardless of the content of the meeting. As mentioned earlier, it is considered that there are three types of individual differences that are likely to occur such as the transcription rate, the grasping rate, and the reduction rate.

Therefore, we will use these three values as indicators of the summarization ability (i.e., listening ability) of the corresponding secretary. Each time the meeting is over, these values are calculated and fed back to the participants in charge of the secretary.

4.4 Discussion Skills Training

Well, in this chapter, to evaluate the discussion ability, we have considered the ability to make statements as a speaking ability, the ability to summarize statements as a listening ability as its constituent elements. Of course, we need to consider other abilities as well, but we focus on discussion ability based on automatically evaluable indices and give feedback to meeting participants quickly. This is to make it possible for each participant to give better results at the next opportunity based on the feedback.

Any training starts with grasping your current state first. And, by being conscious of the current state, the direction of the effort will be decided. Like sports, we improve by accumulating practice. In the case of discussion, at the meeting, the participants make as many statements as possible and make sure that the value of the speaking ability is raised. Also, when the participants are responsible for acting as the secretary, they should act as a secretary with consciousness to increase the value of the indicators of their listening ability.

By the way, our laboratory decides the scores of the seminar based on the data obtained by the discussion mining system (Nagao et al. 2015). However, since we are now thinking of evaluating the process in terms of effort rather than the result, we evaluate the quantity more than the quality of the statements, so it is not always necessary for the outstanding grades to have a particularly high discussion ability. However, the ability of those who are not making efforts will not rise, so people with low scores clearly have low discussion skills.

In order to train the discussion ability and communication ability, it is necessary to record the evaluation results with a considerably long span. Changes in the short-term evaluation results are effective as a clue to evaluate and improve the performance of the developed system, but it will not be enough to judge whether a person certainly has that capability. This is similar to the fact that local optimal solutions do not necessarily become true optimal solutions in optimizing parameters of machine learning models.

It is often said that it takes time for human education, I think that discussion skills as well as basic academic ability need to be firmly acquired based on long-term perspectives. To that end, I think that we must have a clear guide to become a signpost. With clear and unfounded guidance, people will lose confidence in themselves. The technique we developed is useful for clarifying what kind of ability improves what to do. I believe that "evidence-based education" will be possible by such a mechanism.

To the reader, I think that the ability of discussion is a fundamental and important skill for human beings to do intellectual activities. Improving this ability is a task that can be said to be essential for many people. However, if visible growth does not appear, people will get bored with that training. In the next section, I would like to introduce one approach to that problem.

4.5 Gamification for Maintaining Motivation to Raise Discussion Abilities

In recent years, the word gamification has begun to be widely used. Gamification is said to be "movement to incorporate elements of the game into development of social activities and service applications" (McGonigal 2011). Specifically, to introduce various game elements, such as level and reward, into the system to improve the motivation to use the system. The concept of gamification is not new at all, it has been used for various services before. For example, point cards of shops, access ranking of blogs, etc. can be regarded as a type of gamification. In the beginning of the 2010s, it seems that general methods of utilizing such game elements have become generically called gamification.

Many applications for gamification aim to increase the number of times of service used by unspecified number of users and to improve work efficiency, and in most cases, the effect will appear immediately after introduction of gamification. However, in the meetings we are targeting, it is still difficult to improve the discussion ability in a short period of time, although it is possible to facilitate activities within a limited time and temporarily increase motivation. In order for participants to be able to conduct a high-quality discussion, it is necessary to record, evaluate, and support in detail the process of gradually growing and experiencing.

The gamification framework for discussion consists of seven elements as shown in Fig. 4.15. Students unfamiliar with the discussion at the starting point can be motivated by each game element and finally reach the point where their discussion ability has improved.

The elements of this gamification framework are described below.

1. Goal

The purpose of the discussion conducted at the university's laboratory is divided into two categories: deciding future policies by improving the content of research by exchanging views and improving the student's discussion ability.

4.5 Gamification for Maintaining Motivation to Raise Discussion …

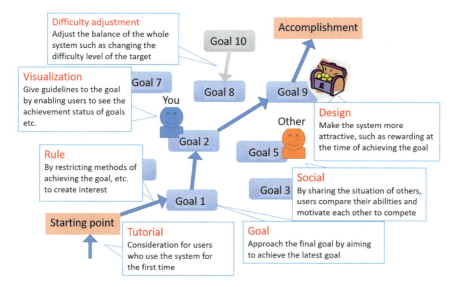

Fig. 4.15 Gamification framework for discussion

Based on the premise that students' abilities will improve, discussion for improving research will be effectively done. We will pay attention to improving student's discussion skills. Therefore, we will make this a final goal.

As a goal element to reach the final goal step by step, it is possible to decompose discussion ability into several components. This decomposed capability has advanced abilities such as "can finally summarize stories" from basic ability such as "can say with loud voice" for example. Meeting participants will voluntarily target the ability they want to improve from the finely disassembled discussion abilities and aim to improve their abilities by speaking to achieve each goal element.

2. Visualization

Meeting participants will be aware of the improvement in capacity and gain a sense of accomplishment by grasping the process of mastering the decomposed discussion abilities independently one by one. For that purpose, it is necessary to have a mechanism to inform the improvement of the discussion ability in real time and a mechanism to confirm the degree of improvement of their own ability later.

3. Rule

As mentioned in the previous sections, when evaluating the discussion ability automatically, semantic evaluation of the contents of the speech is difficult and only quantitative evaluations such as the number of statements and the volume of voice can be done in real time. So, we are making semantic evaluations of the content of each other's statements by participants. Also, instead of evaluating others throughout the discussion, we evaluate each statement at a time aiming to increase the number of statements. Those that regulate the behavior of such participants are called rules.

4. Design

It is a mechanism to improve the motivation for students to participate by attractive interfaces and various compensations. By giving a virtual remuneration, it is necessary to carefully decide factors that become extrinsic motivation, paying attention not to significantly lower the intrinsic motivation.

5. Social

We introduce a mechanism that enables us to check the degree of mastery of discussion skills using Web pages, the users compare their abilities and motivate each other to compete. In addition, it is possible to look back on the statements of other participant's high discussion skills and use them as reference for their own statements. Furthermore, at the end of the meeting, we aim to further motivate competition and satisfy self-esteem desires by ranking and praising participants who demonstrated excellent discussion abilities at that time.

6. Tutorial

For students who join the meeting for the first time belonging to the laboratory, it is necessary to understand not only the discussion mining system described in Chap. 2, but also the elements of the new gamification, which is a heavy burden. Therefore, to understand how to use each system is also one goal in gamification, and we will promote system understanding by presenting detailed guidelines.

7. Difficulty adjustment

The users are roughly divided into two types: teachers and students. Also, from the viewpoint of discussion ability, among students, there is a big difference between students belonging to the laboratory and newly assigned students. By visualizing the difficulty and complexity of the target, participants with a wide range of discussion ability will be able to select appropriate targets for themselves.

We will refer to the mechanism that introduced this gamification framework for discussion as gamified discussion (GD, for short).

The purpose of GD is to keep motivation to improve the discussion skills of students participating in the meeting. In order to evaluate the improvement of the discussion ability, manual evaluation information is required besides the information recorded by discussion mining, but at the same time, by evaluating others during the meeting, the risk of hindering concentration on the discussion.

Therefore, it is necessary to lower the cost of evaluation as much as possible. First, we will break down discussion ability, which is a comprehensive skill of complex manners, so that we can evaluate it more concretely.

For example, the Japan Debate Association classifies abilities necessary for discussion or skills acquired by discussion into the following four categories: (http://www.debate-association.org/).

1. Comprehension ability.
2. Analytical ability.
3. Constitutional ability.

4.5 Gamification for Maintaining Motivation to Raise Discussion …

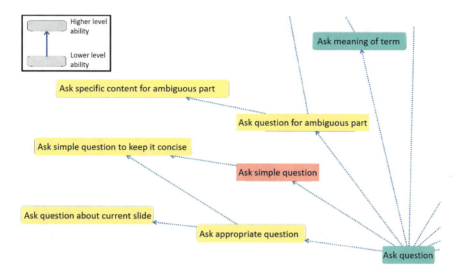

Fig. 4.16 Discussion ability graph (a part)

4. Transmission ability.

Comprehension ability is the ability to understand the background of the agenda, the content and the intention of the statements by the participants. Analytical ability is the ability to analyze whether there is contradiction or ambiguity in the participant's statements after understanding the contents of the discussion. Constitutional ability is the ability to constitute its own statement content so that it is easy to understand and convincing. Transmission ability is the ability to devise ideas for speech, attitudes, gestures, etc. so that the other participants can easily understand.

In addition, participants in the meeting using the discussion mining system must have the ability to reserve the statement and to select the object in the slide by the operation of the discussion commander. In order to conduct a discussion without delay, it is necessary to perform these operations accurately and smoothly, so we decided to classify the ability to deal with systems related to discussion mining as one class of discussion skills. This is called DM ability, and the abilities belonging to the DM ability category such as "I can speak without spending time to reservation" and "speak by referring elements on the screen" have been newly added.

As discussion abilities, there are many inclusion relationships such as "I can say opinions" and "I can say briefly". Therefore, we investigated whether there is an inclusion relation among all the discussion abilities, set the ability to be a subset of the two abilities in inclusion relation as a lower level ability, and make the other the higher level ability. Based on this, we created a graph in which each discussion ability is a node and nodes of two abilities in inclusion relation are linked by links. A part of it is shown in Fig. 4.16. This is called a discussion ability graph. In this graph, discussion ability nodes are color coded for each category.

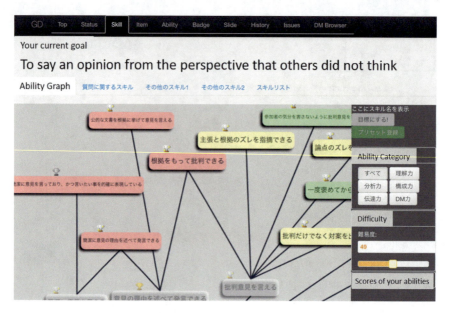

Fig. 4.17 Personal page (goal setting view)

In GD, we hope that finally the comprehensive discussion ability will be improved by acquiring subdivided discussion abilities one by one based on the discussion ability graph. Therefore, we use the process of acquiring each subdivided ability as a goal element in gamification.

The automatic evaluation of the discussion ability described in previous sections was to evaluate whether the way of speaking was appropriate or whether the content being talked was appropriate, but the evaluation of the discussion ability in GD is performed by evaluating the achievement degree of a discussion goal set by one participant by another participant. These are considered to evaluate different aspects of discussion skills. The latter is more difficult to evaluate automatically because there are no clear evaluation criteria. In other words, since the methods based on automatic recognition are suitable for automated evaluation, no human evaluator is necessary, but GD is a method in which multiple participants mutually evaluate their statements during discussion.

In order to minimize the burden on the evaluator, the speaker sets only one of the discussion abilities (goal elements) as a target, while the other participants who are the evaluators listen to the statements and evaluate only with respect to one goal. The meeting participants access the Web page called Personal Page as shown in Fig. 4.17 beforehand and set targets while checking the discussion ability graphs, categories of each ability and acquisition difficulty level.

We also created an evaluation interface to evaluate the speaker's goal during the meeting as shown in Fig. 4.18. The evaluation interface was created as a browser application that can be used with a Web browser so that it can be used from various

4.5 Gamification for Maintaining Motivation to Raise Discussion ...   99

Fig. 4.18 Statement evaluation interface

devices. In particular, it is a user interface that makes it easier to evaluate by touch operation, it is recommended to use a portable device with a built-in touch panel such as a tablet.

Also, when the evaluator determines that the content of the statement is excellent irrespective of the degree of achievement of the target, there are cases where the evaluation of the target includes the evaluation of the quality of the statement. Therefore, a function to evaluate the quality of the statement independently of the speaker's goal was added to the statement evaluation interface.

The evaluator listens to the speaker's statement while checking the speaker's goal displayed on the statement evaluation interface and evaluates whether or not the target has been achieved in five stages. In the five grades evaluation, I made it "it is not done," "it is done a little," "it is half done," "it is done well," and "it is done very well" in order from 1. Evaluation of the quality of the statement makes it possible to input plus 1 or plus 2. This is not to be strictly evaluated in five stages like the goal, but if you think "good," it is positive 1, if you think it is "very good" plus 2. The granularity of these evaluations is the result of adjustment by the authors in order to enable as rapid and precise evaluation as possible.

After entering the evaluation and pressing the submit button (tap for tablet, mouse click for PC) the score will be sent to the server and recorded. At this time, the interface screen switches to the presentation slide display. If the user wants to redo the evaluation, he or she can revise the evaluation of the current statement by displaying the statement evaluation interface again by selecting "Evaluation" from the tab list displayed at the top of the screen.

Fig. 4.19 User information view

When evaluations are made on statements, all the evaluation scores are totaled on the server, and the results are displayed in the statement evaluation interface of the speaker. The final evaluation score for the target of the speaker is the average score of all the evaluations and the total of the evaluation of the quality of the statement. The score for the statement is displayed in the user information view at the upper right of the screen as shown in Fig. 4.19 so that it can be confirmed immediately. In the user information view, only the score for the last speech is displayed.

The speaker can check the evaluation score after the end of the statement and make decisions such as whether to speak again with the same goal or change the goal.

The average of all the evaluation scores with respect to the target goal of the speaker is the point at which the speaker achieves that goal. If the average of the evaluations from all participants is equal to or greater than the "well-done" evaluation, we judge that the speaker has achieved the goal adequately.

At the end of the meeting, participants can check the degree of accomplishment of each discussion ability category on the Personal Page on the Web, while comparing it with other people on the radar chart. As shown in Fig. 4.20, since the current achievement level is visualized easily in graphs, it is possible to set the next goal while grasping their field of strength/weakness. The degree of achievement for each category is the average of all the historical high scores of all targets belonging to that category.

In addition, details of the elements of each category are displayed as a table as shown in Fig. 4.21.

In addition, the Personal Page has a history display function that allows the user to look back at all the past statements of the past meetings, and the user can check the history of statements he or she evaluated in the past and the history of his/her statements evaluated in the past. As shown in Fig. 4.22, for each statement, the contents of the statement and the target goal when he or she speaks are displayed, and when the user press the "Look at the score" button, all evaluation scores for

4.5 Gamification for Maintaining Motivation to Raise Discussion ...   101

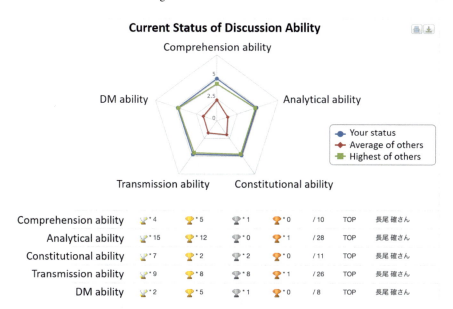

Fig. 4.20 Current status of discussion ability

Fig. 4.21 Details of discussion ability

Fig. 4.22 Past statement data page

that statement are displayed. In order to increase the usefulness of the score while maintaining the anonymity of the evaluator, the evaluator's name is expressed as "teacher" or "student" only. By browsing the history, it is possible to check the difference between the contents of the statement and the target goal and to make a plan to achieve the goal.

In this way, by applying gamification to discussion, the participants can positively improve their discussion abilities. Although the effectiveness of this system has not been fully verified yet, when comparing the average number of statements per unit time of participants in each year before and after GD introduction, it increased from 0.99 to 1.73. In the questionnaire survey, all the participants said that the introduction of GD raised the motivation for discussion. Of course, this alone cannot prove that gamification contributes to improving the discussion ability, but I think that by examining the long-term execution data, we can verify the effect more precisely.

References

R. Barzilay, M. Lapata, Modeling local coherence: an entity-based approach. Comput. Linguist. **34**(1), 1–34 (2008)

K. Kurihara, M. Goto, J. Ogata, Y. Matsusaka, T. Igarashi, Presentation Sensei: A Presentation Training System using Speech and Image Processing, in *Proceeding of ICMI 2007*, pp.358–365 (2007)

J. McGonigal, *Reality Is Broken: Why Games Make Us Better and How They Can Change the World* (Penguin Books, 2011)

K. Nagao, M. P. Tehrani, J. T. B. Fajardo, Tools and evaluation methods for discussion and presentation skills training, SpringerOpen J. Smart Learn. Environ. **2**(5) (2015)

T. Tsuchida, S. Ohira, K. Nagao, Creation of contents and visualization of metadata during face-to-face meetings. Transac. Inf. Process. Soc. Jpn.**51**(2), 404–416, 2010. (in Japanese)

Chapter 5
Smart Learning Environments

Abstract Our university is currently developing an advanced physical–digital learning environment that can train students to enhance their discussion and presentation skills. The environment guarantees an efficient discussion among users with state-of-the-art technologies such as touch panel discussion tables, digital posters, and an interactive wall-sized whiteboard. It includes a data mining system that efficiently records, summarizes, and annotates discussions held inside our facility. We also developed a digital poster authoring tool, a novel tool for creating interactive digital posters displayed using our digital poster presentation system. Evaluation results show the efficiency of using our facilities: the data mining system and the digital poster authoring tool. In addition, our physical–digital learning environment will be further enhanced with a vision system that will detect interactions with the digital poster presentation system and the different discussion tools enabling a more automated skill evaluation and discussion mining. In addition, we argue that students' heart rate (HR) data can be used to effectively evaluate their cognitive performance, specifically the performance in a discussion that consists of several Q&A segments (question-and-answer pairs). We collected HR data during a discussion in real time and generate machine learning models for evaluation. HR data are used to estimate the degree of self-confidence of speech while speakers participate in Q&A sessions. We checked whether there is a correlation between the degree of confidence and the appropriateness of statements. So, we can evaluate the mental appropriateness of the statements. Furthermore, we realized a presentation rehearsal system using virtual reality technology. Based on the system, students can easily experience the act of presentation in front of many audiences with a wide auditorium in virtual space. In this chapter, I will also describe them in detail.

Keywords Leading graduate school · Leaders' saloon · Presentation skills · Digital poster · Psychophysiology-based activity evaluation · Virtual reality presentation rehearsal system

---

© Springer Nature Singapore Pte Ltd. 2019
K. Nagao, *Artificial Intelligence Accelerates Human Learning*,
https://doi.org/10.1007/978-981-13-6175-3_5

5.1 Environments for Evidence-Based Education

We have been developing an advanced physical–digital learning environment that can train students to enhance their discussion and presentation skills. The environment guarantees an efficient discussion among users with state-of-the-art technologies such as touch panel discussion tables, digital posters, and an interactive wall-sized whiteboard. It includes a data mining system that efficiently records, summarizes, and annotates discussions held inside our facility. We also developed a digital poster authoring tool, a novel tool for creating interactive digital posters displayed using our digital poster presentation system. Evaluation results show the efficiency of using our facilities: the data mining system and the digital poster authoring tool. In addition, our physical–digital learning environment will be further enhanced with a vision system that will detect interactions with the digital poster presentation system and the different discussion tools enabling a more automated skill evaluation and discussion mining presented in Chap. 2.

As I mentioned earlier, a lot of attention has been paid to evidence-based research, such as life-logging (Sellen and Whittaker 2010) or big data applications (Armstrong 2014), that proposes techniques to raise the quality of human life by storing and analyzing data of daily activities in large quantities. This technique has been applied in the education sector, but a key method has not been found yet because it is generally hard to record intellectual activities, accumulate, and analyze data in a large scale, and compare it with things like a person's physical activities, position, and movement information. Although there are some recent studies on the automated recording of intellectual activities in more detail, their techniques are not sufficient to be applied to an automated evaluation of a person's intellectual activities. Thus, this study aims to develop a new environment to empower the skills of students not only in real time but also offline based on the abundant presentation and discussion data analyses.

Our study focuses on the new graduate leading program of Nagoya University that aims to cultivate future industrial science leaders. This leading graduate program has a new physical–digital environment for facilitating presentations and discussions among the selected students of the program. In particular, the presentations and discussions of the students are recorded in detail, and the mechanism of knowledge emergence is analyzed based on a discussion mining system. Furthermore, we have evaluated the performance of some students with respect to their skill in creating a digital poster using our recently developed tool.

5.2 Related Work and Motivation

This section has two parts: discussion evaluation and presentation evaluation. For each evaluation system, there are also two parts based on the type of system: a fully automatic system and a semi-automatic system. A fully automatic system calculates

5.2 Related Work and Motivation 107

the scores of discussion or presentation quality in an automated fashion, while a semi-automatic system supports the people in evaluating the discussion or presentation with some evidential data.

5.2.1 Discussion Evaluation

5.2.1.1 Fully Automatic

One of the most familiar types of intellectual and creative activities is discussion at meetings. There is great significance in analyzing discussion in a scientific way and evaluating its quality. Therefore, we proposed and implemented a method for evaluating the discussion ability of students in meetings in a university laboratory setting. There are roughly two kinds of evaluation methods as mentioned in Chap. 4:

(1) One based on acoustic information, that is, the manner of speaking.
(2) One based on language information, that is, the contents of speech.

Method (1) evaluates the appropriateness of utterances in a discussion by using the acoustic characteristics of speech. The characteristics are automatically evaluated in real time and fed back to speakers during a meeting. For example, we measure the voice size (loudness), voice intonation, speech speed, fluency, tempo, and other vocal aspects of a speaker and automatically evaluate the acoustic appropriateness of the statements. If anything is determined to be inappropriate, the system provides feedback to the speaker in several ways, such as with a message popping up on a screen.

Method (2) analyzes linguistic characteristics in consideration of context. For example, we estimate the consistency of the context of statements by using machine learning techniques. Then, the linguistic appropriateness of the statements is automatically evaluated.

We believe that carefully examining these methods over a period of time will result in a more detailed analysis that helps us focus on more appropriate training for students.

Students' improvement in discussion ability is evaluated in subsequent training. Discussion–skill training is carried out through a repeat cycle that consists of notifying a person of a problem and giving advice via e-mail prior to a meeting, evaluating statements during the meeting, and the person reflecting and making improvements after the meeting.

This section reviews other methods for discussion evaluation including online discussions such as text discussion on bulletin boards.

With the abundant data in discussions, there is difficulty in searching for good quality posts. An automatic rating of postings in online discussion forums was presented based on a set of metrics (Wanas et al. 2008). This set of metrics was used to assess the value of a post and includes the following: relevance, originality, forum-specific features, surface features, and posting-component features. With these metrics used to train a nonlinear support vector machine classifier, the posts were then categorized to their corresponding levels (High, Medium, or Low).

Another system called Auto-Assessor used natural language processing tools to assess the responses of students to short-answer questions (Cutrone et al. 2011). The system utilized a component-based architecture with a text preprocessing phase and a word/synonym matching phase to automate the marking process. In their system evaluation, they compared the assessment results of the Auto-Assessor and Human Graders to verify the possibility of applying the proposed system in practical situations.

However, these fully automatic systems still have some drawbacks. Some methods are language independent resulting in poor performance in relevance and originality (Wanas et al. 2008); thus, other additional techniques should be employed in their assessment of discussions. Also, even with additional NLP techniques, the weights given to words are not varied (Cutrone et al. 2011) hindering the system from identifying words that are more significant than others.

5.2.1.2 Semi-automatic

Aside from fully automatic systems, some studies employed a semi-automatic approach. One such study is the implementation of a group discussion evaluation method and a discussion evaluation support system that focused on ex-post evaluation (Omori et al. 2006). The system provided a Web-based interface to display the evaluation item and the evaluation criteria so that users can easily make a score to each of the discussion remarks based on clearness of remarks, proposal of issues, and logicality of remarks. Results confirmed the effectiveness of both their evaluation method and support system.

With the abovementioned systems, there was no mention about one problem in discussions, which is the difficulty in getting students to actively participate. Thus, a gamification framework was integrated into a discussion support system for enhancing and sustaining motivation in student discussions (Ohira et al. 2014). Besides sustaining student motivation, the system also evaluates and visualizes improvement of the students' capacity to discuss. It also supports the users to evaluate the quality of each discussion statement.

However, with the two semi-automatic systems, more experiments are needed to determine the effect of teachers' feedback to the students (Omori et al. 2006) and its performance in real-world settings.

5.2 Related Work and Motivation

5.2.2 Presentation Evaluation

5.2.2.1 Fully Automatic

A presentation training system called Presentation Sensei was implemented to observe a presentation rehearsal and give presentation feedback to the speaker (Kurihara et al. 2007). The system is equipped with a microphone and camera to analyze the presentation by combining speech and image processing techniques. Based on the results of the analysis, the system provides the speaker with recommendations for improving presentation delivery such as speed and audience engagement. During the presentation, the system can alert the speaker when some of the indices, speaking rate, eye contact with the audience, and timing, exceed predefined warning thresholds. After the presentation, the system generates summaries of the analysis results for the user's self-examination. Although this system focuses on self-training, it still needs to be tested in a real presentation environment.

5.2.2.2 Semi-automatic

Another presentation training system called PitchPerfect was implemented to develop confidence in presentations (Trinh et al. 2014). From interviews with presenters, the authors uncovered mismatches between best rehearsal practices as recommended in the presentation literature, the actual rehearsal practices, and support for rehearsal in conventional presentation tool. Thus, they developed the proposed system, an integrated rehearsal environment that supports users to evaluate their presentation performance during preparation for structured presentation in PowerPoint. Their user study with 12 participants demonstrated that PitchPerfect led to small but significant improvements in perceived presentation quality and coverage of prepared content after a single hour of use, arising from more effective support for the presenter's content mastery, time management, and confidence building.

5.2.3 Motivation

In the initial phase of our research, we selected a semi-automatic approach to evaluate the discussion and presentation. However, our proposed system can acquire several kinds of student activity-related data to make evaluation automated in the near future. We understand that current technologies that analyze human activity data which are fully automated are still insufficient to realize our purposes so we focused on data acquisition by using our new environment for discussion and presentation.

Fig. 5.1 Leaders' saloon environment

5.3 Leaders' Saloon: A New Physical–Digital Learning Environment

The Leaders' Saloon shown in Fig. 5.1 is capable of creating discussion contents using discussion tables, digital poster panels, and an interactive wall-sized whiteboard.

5.3.1 Discussion Table

Each student uses a tablet to connect with the facilities including the discussion table shown in Fig. 5.2. The content and operation history of the discussion table are automatically transferred and shared to the server, the meeting cloud. Previous table content can be easily retrieved, and any texts or images can be reused. Such reference and quotation operations are recorded and analyzed to discover semantic relationships between discussions. Furthermore, software that analyzes temporal changes of table contents with the corresponding users is also being developed.

5.3.2 Digital Poster Panel

For poster presentations, a digital poster panel system, shown in Fig. 5.3, is used for content and operation analyses. The system helps the users create digital posters and analyze their creation process. The system also supports the retrieval of previously presented posters and allows users to annotate them. Annotations are automatically sent to the author and are analyzed by the system to evaluate the quality of the poster. Poster presentations as well as the regular slide-based presentations are also broadcasted by streaming on the Web. The system collects and analyzes the feedbacks

5.3 Leaders' Saloon: A New Physical–Digital Learning Environment

Fig. 5.2 Students using discussion table

based on comments and reviews given by Internet viewers (e.g., Twitter users can associate their tweet messages with any scenes from the presentation based on the starting and ending timestamps).

5.3.3 Interactive Wall-Sized Whiteboard

As shown in Fig. 5.4, our facility houses a wall-sized whiteboard. Unlike the traditional whiteboards, we are able to physically and digitally write on the whiteboard. We use a special projector equipped with an infrared sensor to detect the location of the digital pen with respect to the wall. The writings and interaction on the whiteboard can then be recorded by cameras. The captured data using the camera can identify the physical interaction in combination with the given digital interaction information. This system is under development, and we are working on proposing a new evaluation system that can enhance the presentation and discussion performance of students using this system.

Fig. 5.3 Poster presentation using digital poster panel system

5.4 Importing Discussion Mining System into Leaders Saloon

We developed an extended version of the discussion mining system presented in Chap. 2 working at the Leaders' Saloon. The discussion tables are used to operate and visualize discussion structures. The users also use discussion commanders and the previously described discussion mining system.

In this section, we explain two systems implemented on the discussion tables to visualize real information recorded by the discussion mining system: (1) discussion visualizer, a system to visualize the structure of an ongoing discussion, and (2) discussion reminder, a system to retrieve and visualize past discussions.

5.4.1 Discussion Visualizer

The discussion visualizer shown in Fig. 5.5 is a system to visualize the structure of meeting discussions shown in the discussion table (Sect. 3.1). This visualizer consists of a meeting view, a slide list, a discussion segment view, and a discussion segment list.

5.4 Importing Discussion Mining System into Leaders Saloon

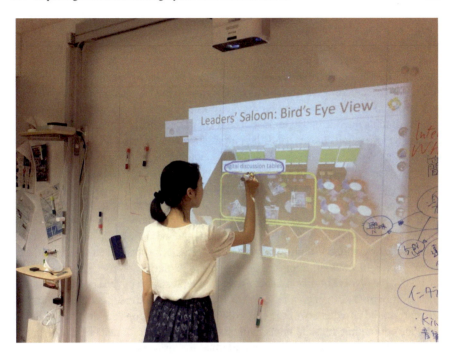

Fig. 5.4 Capturing data using interactive wall-sized whiteboard

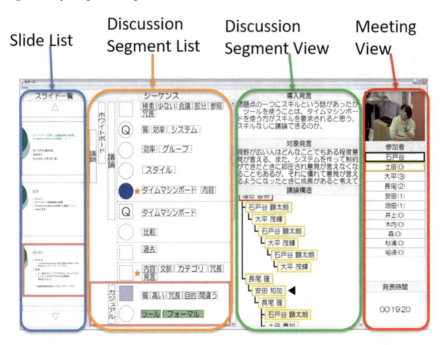

Fig. 5.5 Discussion visualizer

The meeting view provides a preview of camera records showing the participants, a list of all attendances, and the elapsed time of the presentation. A list of slide thumbnails displayed in the presentation are also shown, and the thumbnail of the currently displayed slide is emphasized in the slide list. Speakers can operate the slideshow by selecting the thumbnail in this view using the touch panel.

The discussion segment view shows the information about the discussion segment, which contains the current statement. The texts of the start-up statement, which was the trigger of the discussion, and the parent statement of the current statement (if it is a follow-up statement) are shown at the upper side of this view. The structure of the discussion segment is shown at the bottom side of this view. Users can also make corrections of parent statements. Participants confirm the stream of discussion at the meeting through the discussion segment list. In this list, the nodes representing main topics are shown as rectangle nodes while the subtopics are shown as circle nodes. These discussion segment topics are displayed as a chain structure in the middle, the keywords of multiple discussion segments are displayed on the left, and the keywords of the main topics or subtopics are displayed on the right. Moreover, the nodes that involve questions and answers are represented by the specific character Q. The amount of agreements on the statements input by the discussion commanders are represented as a density of the color of the nodes. The icons are displayed next to the node containing the statements marked by discussion commanders. Therefore, it enables participants to confirm when important discussions occur.

There are various kinds of discussion segments created by the discussion mining system, for example, short segments with only comments on the presentation and long segments that contain a lot of statements resulting from intense debate. There is also a possibility that long discussion segments have follow-up statements whose content derives from the topic of the start statement. Thus, we think that the start statement is the root node of the discussion segment and some subtopics derive from this root node.

5.4.2 Discussion Reminder

A review and sharing of previous discussion contents lead to a uniformed knowledge level among all participants, wherein low-level participants can make remarks actively. This will also prevent redundant discussion. From here, we can then think about topics from a new point of view and figure out solutions to problems that have not been solved due to lack of technology. Therefore, we develop a system to retrieve and browse past discussions on time, called the discussion reminder.

There are two concerning issues in the development of the discussion reminder. One is an accurate understanding of discussion contents, and another issue is the quick retrieval of discussion contents preventing any disruption in the ongoing discussion. Unclear and inadequate sharing of discussion content will inhibit the achievement of a uniformed knowledge level and will lead to misunderstandings and confusion. Thus,

5.4 Importing Discussion Mining System into Leaders Saloon

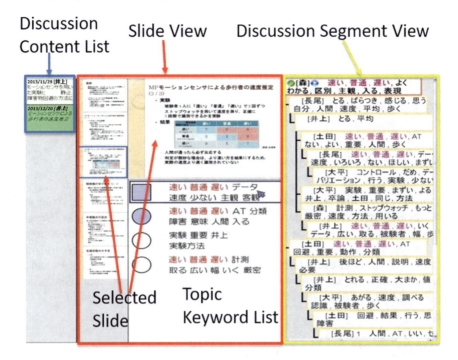

Fig. 5.6 Discussion reminder

the discussion reminder provides a function to browse videos of past discussions for accurate understanding.

However, all participants need to interrupt the ongoing discussion for a review of discussion content, and thus it is desirable to finish the review in no time using the above method to find the things required in the audiovisual information. For an efficient review, the discussion reminder provides an interface to narrow down the browsing information, such as discussion content matched with queries, slides matching discussion content, and statements associated with matched slides, and to cooperative retrieval by participants. A participant who notices the existence of the discussion, which he/she wants to review, inputs queries to the discussion reminder. Various types of information, such as names of presenters, dates of meetings, and keywords, are available as queries. The contents of retrieved results are displayed on the discussion table as shown in Fig. 5.6.

Participants conduct various operations using the touch panel in this interface. This interface consists of a discussion content list, a slide list, and a discussion segment view. The discussion content list displays titles of the discussion contents, which contains the discussion matched queries. When a participant selects a title using the touch panel, slide thumbnails comprised in the selected discussion content are shown at the bottom of the slide list. Participants can preview the larger slide thumbnail at the top of the slide list.

The discussion segment view shows information about the discussion segments associated with the slide selected in the slide list. Examples of discussion segment information include structures of discussion segments, speaker's ID, keywords of statement, and so on. In the discussion segment view, full text of the statement can also be previewed. Participants can browse videos in the video view displayed on the table from the start time of the selected statement in the discussion segment view.

5.4.3 Employing Machine Learning Techniques

In this study, machine learning techniques are employed to obtain deep structures of presentation and discussion content. Techniques like deep neural networks (Bengio 2009) integrate several contexts of information such as operation histories of users. By integrating the results of subject experiments on presentations and discussions, different methods to evaluate the quality of students' intellectual activities and to increase their skills are discovered. The system tries to perform some consensus-building processes to make evaluation results appropriate for each student.

5.5 Digital Poster Presentation System

The digital poster presentation system consists of an authoring tool for digital posters, an interactive presentation system with digital posters, and an online sharing system for digital posters. Poster presentations can be considered as a close communication with the audience, and are also ideal for training in discussion not only for presentation. The digital poster presentation system makes the poster presentation easier. Tools such as PowerPoint slides can be integrated into the poster presentation. Additionally, the system will be extended for an interactive data acquisition. Hence, we believe that this system would significantly change the way of poster presentations.

5.5.1 Digital Posters Versus Regular Posters

A digital poster is an interactive multimodal version of regular papers. The advantage of digital posters includes retrieval and reuse of content. However, one of the biggest problems is portability since a digital poster needs special hardware such as a digital poster panel and these devices cannot be carried elsewhere. Perhaps, in the near future, large and thin film-type screen devices, such as organic electroluminescence displays, will be available and tools for digital posters will be commodities and easily acquired.

5.5 Digital Poster Presentation System

Fig. 5.7 Main screen of digital poster authoring tool

5.5.2 Authoring Digital Posters

Authoring of digital posters is very simple but some preparation is needed. The users should prepare resources such as images, videos, and slides in advance. We also developed an online resource management system for memos, images, videos, and slides. The digital poster authoring tool can import any resources submitted or shared in the resource management system.

The digital poster authoring tool shown in Fig. 5.7 has three parts: the main menu, the resources menu, and the poster field.

The main menu provides the basic functionalities of the tool such as creating, opening, and saving of poster files, setting up the desired preferences, and choosing different creation modes. The digital authoring tool is also able to create both portrait and landscape orientation posters as needed.

The resources menu shown in Fig. 5.8 lets the users add different types of blocks to the poster field. Each block automatically downloads a certain type of resource depending on the selected block from the online resource management system, except for the layout and text blocks. Selecting an image block will automatically scan for images in the resource management system, while selecting a video block will automatically scan for videos in the resource management system. For the slide block, existing PowerPoint slides will be selected.

When the user taps a block in the resource menu, a list of thumbnail images is displayed in the window that appears from the right edge of the screen as shown in Fig. 5.9. The user can easily arrange the layout of the poster using a layout block and interactively change a position of a block's borderline. When the user wants to place any resource in the block, he/she should just drag and drop the thumbnail image from the resource list to the target block as shown in Fig. 5.10.

118   5 Smart Learning Environments

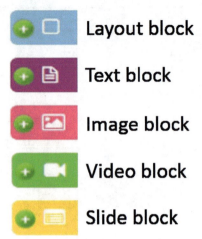

Fig. 5.8 Resource menu

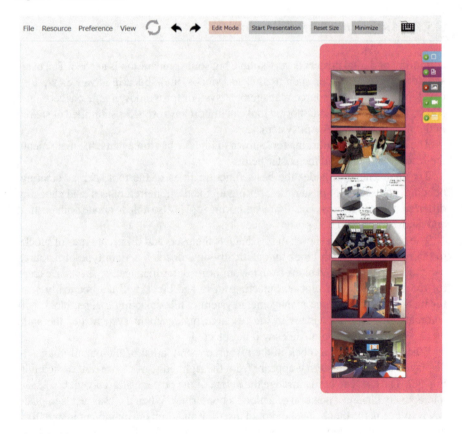

Fig. 5.9 Image resource menu

5.5 Digital Poster Presentation System

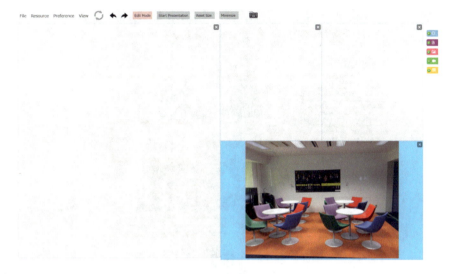

Fig. 5.10 Image resource placement in layout block

Other resources, such as videos and slides, are inserted in the blocks in a similar way. An example of a created poster using the described authoring tool is shown in Fig. 5.11. When the user finished editing the digital poster, the final poster can be stored in the online poster sharing system. It can be used for presentations by searching the digital poster at any time. During presentation time, the enlargement of images and the playback of videos and slides in the poster can be done.

5.5.3 Data Acquisition from Interactions with Digital Posters

Digital posters are not only for a presenter to make a presentation but also for an audience to view in detail by interacting with the poster. Posters are unlike slides, where the complete content is summarized in one piece, which is more suitable to understand the content quickly. At the Leaders' Saloon, visitors can easily retrieve and view the digital posters using the digital poster panel whenever they like. Interaction histories when visitors have interacted with the posters are recorded automatically. The number and time of poster views, views of the elements in the poster, and data such as browsing the order of the poster elements can be obtained by this system. These data are used to evaluate the posters and the skills of the poster author.

Fig. 5.11 Example of digital poster

5.6 Skill Evaluation Methods

The focus of this study is the students of the new Graduate Leading Program at Nagoya University, which aims to nurture future global leaders. To achieve this goal, improving the communication skills of the students must be addressed. In this study, we focus on developing the discussion and presentation skills of the students and this section describes in detail the evaluation method for the discussion and presentation skills of the students.

Student	A	B	C	D	E	F	G	H	I	J	K	L
(1) Participation	33	34	36	36	35	34	36	33	36	36	36	36
(2) Presentation	3	3	2	2	3	3	3	3	3	3	3	3
(3) Secretary	3	3	3	3	3	3	3	3	3	3	3	3
(4) Start-up	39	47	34	54	53	34	49	50	43	47	48	68
(5) Start-up except Presenter's cases	39	44	34	54	51	34	49	48	43	47	48	68
(6) Follow-up	113	151	113	120	127	169	199	141	162	165	149	194
(7) Follow-up except Presenter's cases	31	25	35	37	19	69	55	35	50	42	40	48
(8) Quality	26	27	40	42	63	58	76	92	102	112	125	145
(9) Quality per Statement	0.17	0.13	0.27	0.24	0.35	0.28	0.3	0.48	0.49	0.52	0.63	0.55
Score (1)x2+(2)x10+(3)X5 +(5)x3+(7)x2+(8)x4	394	403	439	511	558	585	678	693	754	790	841	997

Fig. 5.12 Data results of discussion mining system

5.6.1 Discussion Skills

I presented discussion skills evaluation methods based on acoustic and linguistic features of participants' statements in Chap. 4. In the Leaders' Saloon, we employed a simpler version of the evaluation method based on statistic data calculated by the discussion mining system.

Data acquired by the discussion mining system includes participant types (presenter, secretary, and others), number of start-up/follow-up statements of each participant, and quality scores for each statement. The quality scores are calculated by the agreement/disagreement data input by each participant's discussion commander during discussions. For each statement, one point is added if someone agrees with it, one point is subtracted if someone disagrees, and then the score is determined. Results of the aggregate data of multiple students in 3 months are shown in Fig. 5.12.

The discussion skills of a student are evaluated using the score calculated by the following processes. First, the weight values for every behavior are determined. These weights are going to be rationally determined in the future using machine learning, but for now, the values were decided intuitively based on the difficulty of execution.

- Number of participants: 3.
- Number of presentations: 10.
- Number of secretary acts: 5.
- Start-up statements except presenter's cases: 3.
- Follow-up statements except presenter's cases: 2.
- Quality (sum of agreement/disagreement values): 4.

Let the score be the value of the sum after having applied such weight to the number of each behavior. Additionally, the evaluation of statement quality is also calculated. For the discussion skill score calculation, the presenter's cases are excluded by start-up and follow-up statements of the students. This is because the situations when the presenters must answer the question from other participants occur naturally and they should not be treated the same way as cases in which participants of the discussion make remarks spontaneously.

The students can judge their status for this evaluation as a reference and can analyze their weak points. The student performance increases if the student makes many statements when he/she is not the presenter, and many of the other participants also agree with these statements. It is then possible that these data be a basis to improve a student's discussion skills. It can also be confirmed that discussion skills are improved by making high-quality statements, that is, a lot of agreements obtained from many participants.

5.6.2 Presentation Skills

A study on developing oral presentations skills embedded oral presentations and assessment to their curriculum (Kerby and Romine 2009). In their case study, they included at least one oral presentation in three of their courses and used a rubric to assess the oral presentations. Their results indicate that students better understood their weaknesses, strengths, and areas for improvement with their presentations. In our study, we also implemented the same design to improve the presentation skills of our students. We conducted two poster presentation sessions with two groups of students to evaluate their presentation skills. We used the poster presentation format instead of the regular oral presentations because of the interactivity of poster presentations. In poster presentations, the students are able to engage in conversation with people, giving them more opportunities to improve their communication skills. Also, poster presentations enliven the student presentations because students interact with each other more instead of just passively observing like in formal presentations.

In the case of oral presentations, we developed a presentation training system in a virtual reality environment. The system also works at the Leaders' Saloon. I will explain the system at the end of this chapter.

5.6.2.1 Poster Presentation Session I

In this poster presentation session, twenty-four (24) inexperienced students were divided into six (6) groups and each group was asked to create a digital poster using the authoring tool discussed in Sect. 5.5. Each group presented their posters in the allotted time of fifteen (15) minutes, while members of the other groups and spectators of the poster session evaluated each poster presentation. The evaluation sheet used for this poster presentation session is shown in Fig. 5.13.

Evaluators fill up the feedback form shown in Fig. 5.13 for each presentation. The evaluation criteria include Content, Organization, and Impact. The said criteria are based on the common themes in Brownlie's 2007 bibliography (Hess et al. 2009). For scoring results, the numerical values for the different ratings are as follows: Bad

5.6 Skill Evaluation Methods

Criteria	Bad	Poor	Fair	Good	Excellent
Content					
Clarity of content					
Quality of content Is it complete with sufficient background, methodology, and results?					
Supports main points					
Organization					
Layout Is it logically organized into sections with text and graphics that flow from one part to the next?					
Appropriate font size Is it readable?					
Important information is readily available and easy to grasp					
Clearly identified topic and purpose Does it emphasize the most important components of the research?					
Informative and clear Well-chosen graphics and use of color to emphasize key points					
Impact					
Did it catch your interest?					

Fig. 5.13 Poster presentation evaluation sheet I

Criteria \ Groups	I	II	III	IV	V	VI
Content						
Clarity of content	3.4444	2.8333	3.7600	3.6364	3.6250	3.3333
Quality of content	3.4828	2.6667	4.0000	3.5833	3.7200	3.5769
Supports main points	3.5200	2.6957	3.8696	3.7273	3.5000	3.3182
Organization						
Layout	3.1724	2.7778	3.7778	3.6400	3.4615	3.4615
Appropriate font size	3.4483	3.1111	3.5185	3.5600	3.3846	3.1154
Important information is readily available and easy to grasp	3.4286	2.7692	3.6538	3.5000	3.4800	3.5000
Clearly identified topic and purpose	3.4828	2.5556	3.6667	3.4000	3.4615	3.6000
Informative and clear	3.3793	2.7778	3.8148	3.4400	3.6400	3.4400
Impact						
Did it catch your interest?	3.1379	2.9231	3.8519	3.4000	3.5833	3.3846

Fig. 5.14 Poster presentation I score results

is 1, Poor is 2, Fair is 3, Good is 4, and Excellent is 5. The average scores of all the evaluators for all group presentations are shown in Fig. 5.14.

In addition, the scores and standard deviations of all groups are shown in Fig. 5.15. Since the standard deviations are not too large, this metric is not far off from human intuition for evaluation. Histories of interactions with digital posters are not analyzed yet. We are planning to combine human metrics and accumulated data such as access counts of posters and their internal elements in the near future.

However, there were major drawbacks in the evaluation sheet that we used for this session. First, we failed in evaluating the presentation delivery of the students. Second, the evaluators found it hard to judge certain criteria based on the ratings. Thus, for the next poster presentation session, we made a number of changes in our evaluation sheet.

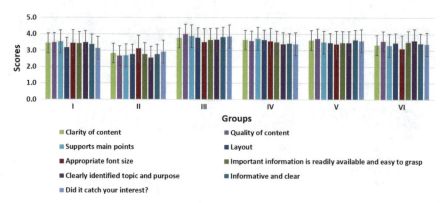

Fig. 5.15 Scores and standard deviations of all groups

5.6.2.2 Poster Presentation Session II

In this poster presentation session, five (5) students were asked to create and present five digital posters using the authoring tool discussed earlier. Spectators of the poster session were asked to evaluate each poster presentation.

We improved our evaluation criteria based on the encountered problems in the former trial. Based on feedback from the evaluators, one major drawback in the previous evaluation sheet shown in Fig. 5.13 is that there are not enough details for the ratings (Bad, Poor, Fair, Good, and Excellent) under certain criteria. Thus, the evaluation sheet was modified to a rubric with concrete descriptions for each score and criteria. The new evaluation sheet is shown in Fig. 5.16. Using this rubric style of evaluation, the evaluation criteria were clearer and evaluation time were faster for the spectators. Aside from changing the evaluation format, the sets of criteria were also modified. We added two main sets of criteria: Impact and Presentation.

We added Impact to determine how the poster is able to attract the attention of spectators. It consists of the criteria for evaluating the poster's title, overall appearance, and interest. We also added Presentation, another set of criteria for evaluating the students' poster presentation skills. It consists of the ability to communicate properly to their audience and the ability to answer questions confidently. Adding these new sets of criteria provided a more effective and complete digital poster evaluation. Using the new evaluation sheet, the score results of the second round of poster presentations are shown in Fig. 5.17. The evaluation scores of the professors (P) and students (S) were calculated. With these results, we were able to determine the weakness of each poster based on the criteria. For example, with Poster III, its content and organization needed a lot of improvement thus for feedback, the author needs to focus on these sets of criteria when he/she modifies the said poster.

Criteria	0	1	2	Score
Impact				
Title	None	Unclear	Concise and stands out	
Overall Appearance	Not visually appealing, no color or graphics	Uses colors and graphics	Visually appealing, nice colors and graphics	
Interest	Not noticeable	Noticed but not interested	Caught my interest	
Content				
Quality	Has no background, methodology, and results	Insufficient background, methodology, and results	Complete and sufficient background, methodology, and results	
Main Points	Not presented	Presented but not obvious, may be embedded in the text / graphics.	Explicitly labeled (e.g. Objectives, Conclusion, Results)	
Organization				
Logical Structure	Cannot figure out how to move through the poster	Headings imply the flow	Very easy to navigate through the content	
Text Size	Too small	Readable	Very easy to read	
Text/Graphics Balance	Too much text or too much graphics	Not enough text to describe the graphics	Balanced text and graphics, enough text to explain the graphics	
Presentation				
Communication	Voice is inaudible and explanation is not understood	Explanation is somewhat unclear	Able to explain content well and clearly	
Questions	Answering of questions is lacking	Answering of questions is inadequate	Able to answer questions clearly	

Fig. 5.16 Poster presentation evaluation sheet II

5.7 Future Features of Smart Learning Environments

The current training environment contains a 2D interactive system, such as touch panel discussion tables, digital poster panels, and an interactive wall-sized whiteboard, facilitating the interactions of users with the system. However, to further enhance the performance of the current learning environment, a vision system will be incorporated to increase the interaction dimension to 3D. The system will consist of a multi-camera system or Kinect that has a camera and range sensor device. Moreover, given an intelligent system that recognizes the users by robust face detection algorithm, user interaction will be smooth, and annotations will be automated and

Poster		I		II		III		IV		V	
Criteria/Evaluators	Professors	Students	Professors	Students	Professors	Students	Professors	Students	Professors	Students
Impact										
Title	1.6000	1.2500	1.5000	1.7000	1.4000	1.8571	1.0000	0.9091	1.3333	1.7500
Overall appearance	1.0000	1.0833	1.3333	1.4000	1.2000	1.0000	1.2500	1.3636	1.6667	1.2500
Interest	1.4000	1.4167	1.3333	1.4000	1.2000	1.2857	1.2500	1.4545	1.5000	1.2500
Content										
Quality	1.0000	0.9167	1.1667	1.3000	0.8000	1.0000	1.2500	1.6364	1.1667	1.0000
Main points	1.2000	1.0833	1.1667	1.5000	0.6000	0.8571	1.2500	1.5455	1.3333	1.2500
Organization										
Logical structure	1.0000	1.1667	1.1667	1.5000	0.6000	0.8571	1.0000	1.2727	1.3333	1.3750
Text size	1.2000	1.0000	0.8333	1.2000	0.8000	1.0000	1.2500	1.0000	1.5000	1.1250
Text/graphics balance	1.4000	1.4167	0.8333	1.5000	0.6000	0.7143	1.5000	1.4545	1.5000	1.5000
Presentation										
Communication	2.0000	1.4167	1.5000	1.0000	1.2000	1.0000	2.0000	1.4545	1.5000	1.5000
Questions	1.8000	1.5000	1.5000	1.4000	1.2000	1.0000	1.2500	1.4545	1.5000	1.2500
Total Score	**13.6000**	**12.2500**	**12.3333**	**13.9000**	**9.6000**	**10.5714**	**13.0000**	**13.5455**	**14.3333**	**13.2500**

Fig. 5.17 Poster presentation II score results

personalized, thereby creating a more advanced learning environment. We plan to utilize an automated evaluation system and facilitation system for intellectual activities to determine whether students' skills are improving and whether newly created content is more highly evaluated than previous content.

The current criteria do not evaluate things like body movement, gestures, and posture, which is common in evaluating presentations. However, it is difficult to evaluate these criteria and we will be incorporating evaluating these movements through the planned vision system (the virtual reality system mentioned later has a recognition function of body movement and gestures). In addition, students' psychophysiological data such as heart rate will also be incorporated in our learning evaluation system. A trial system is explained next. In order to improve the evaluation criteria presented in the previous section, the automated system is expected to receive real-time evaluation from the audience and to provide the presenter with the relative score. The registered audience can input their score to an online sheet with a tablet while attending the poster session. The audio–visual system is also expected to record the visual and audio interactions between the presenter and each audience. The system will be able to match the provided online score by each spectator and his/her interaction with the presenter. We hope to understand the relation between the audio/visual information and the provided online score to train the system. Our eventual goal is to introduce a nearly automatic evaluation system that is regularly trained by the audience.

A novel physical–digital learning environment for discussion and presentation skills training has been developed at our university under the leading graduate program. By using state-of-the-art technologies, the selected students of the program will achieve an effective, interactive, and smooth discussion with the discussion mining system simultaneously summarizing and annotating the ongoing discussion. The discussion contents are available to the community or to the faculty for evaluation, feedback, and follow-up activities. With this prototype environment, a new education system may emerge promoting efficient and advanced learning.

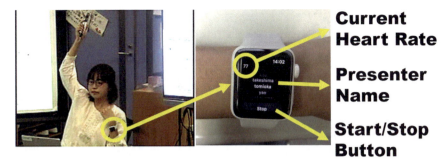

Fig. 5.18 Heart rate acquisition system based on Apple Watch

5.7.1 Psychophysiology-Based Activity Evaluation

Considering that the discussion process is a type of cognitive activity, which could result in changes in certain psychophysiological data, such as heart rate (HR) variability (HRV), several studies have proven that HR is an important index of the autonomic nervous system regulation of the cardiovascular system (Camm et al. 1996). Therefore, there has been increasing focus on observing the correlation between HR data and cognitive activities. A study on measuring the HR during three cognitive tasks (Luque-Casado et al. 2013) revealed the affection of cognitive processing on HRV. The stress level also has been assessed during Trier social stress test tasks, a type of cognitive activity, by using HR and HRV metrics (Pereira et al. 2017). Judging from the large amount of evidence presented, we argue that the HR data of the participants of a meeting can be used to effectively evaluate the answer quality of Q&A sessions in the meeting, which is helpful in improving participants' discussion skills (Peng and Nagao 2018).

Smart watches, such as Apple Watch, the Fitbit series, and Microsoft Bands, contain wearable sensors to accurately detect users' biological data, such as HR and blood pressure. Such noninvasive detection makes it possible to link users' biological information with their daily activities. Iakovakis and Hadjileontiadis (2016) used the Microsoft Band 2 to acquire the HR data of users to predict their body postures. In our study, we used the Apple Watch to collect participants' HR data on the basis of our DM system and to visualize the data during discussions. Through the Health Kit framework on the Apple Watch, which we asked participants to wear on their left wrist during discussions, participants' HR data were acquired almost in real time in 5–7 s intervals, as shown in Fig. 5.18. The collected HR and participants' information is displayed on the Apple Watch screen as well as synchronously presented on an HR browser.

To automatically evaluate the discussion performance, we started from analyzing the answer quality of Q&A segments in discussion chunks mentioned in Chap. 2, which are the most important constituent components generated around a discussion topic. Our goal was to validate our argument that the HR of discussion participants

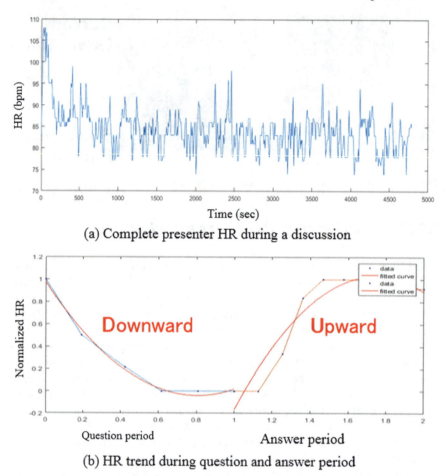

Fig. 5.19 Heart rate graphs

can be used to effectively evaluate the answer quality of Q&A segments during discussions.

All HR information of participants during their discussions is displayed in a graph, shown in Fig. 5.19a, that presents a participant's complete HR detected per minute throughout a discussion. The HR segments in each Q&A segment were then extracted and displayed in a graph, shown in Fig. 5.19b, which shows HR data during a question period (blue line) and answer period (orange line). We then computed 18 h and HRV features from all Q&A segments as well as the question-and-answer periods separately.

The HR and HRV features include mean, standard deviation (std.), and root mean square successive difference (RMSSD) from these two periods (question-and-answer periods), and these metrics have been proven to be important for understanding HRV

Fig. 5.20 HR and HRV features

HR period	HR and HRV features
Both periods	mean, std., RMSSD, trend, freq. all mean, freq. all std.
Question period	mean, std., RMSSD, trend, freq. question mean, freq. question std.
Answer period	mean, std., RMSSD, trend, freq. answer mean, freq. answer std.

differences under cognitive activities (Wang et al. 2009). The trends in the HR of these two periods are also computed by calculating the difference between two adjacent HR points. If the number of positive differences was more than the negative ones, we assumed that the HR period showed an upward trend; if not, it showed a downward trend, as shown in Fig. 5.19b. We used a quadratic curve (red line) to more clearly present the HR trend. We can see that HR during the question period showed a downward trend and upward trend during the answer period.

We also divided the HR data of these two periods into nine ranges: less than 60, 60–70, 71–80, 81–90, 91–100, 101–110, 111–120, 121–130, and more than 130 bpm. The mean and std. were calculated to describe the HR appearance-frequency distribution in each range. Figure 5.20 summarizes these 18 features.

We collected discussion data from nine presenters from nine lab-seminar discussions held over a period of 4 months. Twelve undergraduate and graduate students and three professors made up the participants. The discussions were carried out following the presenters' research reports, with the participants asking questions related to the discussion topic that were then answered by the presenters. There were 117 complete Q&A segments extracted from these 9 discussions, and the answer quality of these Q&A segments was evaluated by the participants who asked the questions by giving a score based on a five-point scale: $1 =$ very poor, $2 =$ poor, $3 =$ acceptable, $4 =$ good, and $5 =$ very good. We obtained 66 high-quality answers with scores from 4 to 5 and 51 low-quality answers with scores from 1 to 3.

We adopted three machine learning models, logistic regression (LR), support vector machine (SVM), and random forest (RF), to carry out binary classification of the Q&A segments' answer quality. About 80% of Q&A segments were randomly selected as a training dataset and the remaining 20% as a test dataset.

For the LR model, we obtained a 0.790 F-measure by using an eight-feature candidate subset and an F-measure of 0.740 by using a seven-feature candidate subset; therefore, we used the eight-feature subset to train our LR model. We obtained an F-measure of 0.805 for the SVM model with 10 h and HRV features we selected in advance. For the RF model, when there were 36 trees (submodels of RF) and 19 terminal nodes on each tree, we obtained the highest F-measure of 0.87. In this case,

Fig. 5.21 Evaluation results
of each learning model

Evaluation model	F-measure
LR	0.790
SVM	0.805
RF	0.870

we chose an eight-feature subset. Figure 5.21 lists the evaluation results for each model.

Comparing the F-measures of each model, the RF model exhibited superior evaluation performance compared to the LR and SVM models. Considering all three models, the HRV data of participants showed an outstanding performance in evaluating Q&A segments' answer quality. Meanwhile, we focused on seven HRV features: all mean, answer trend, all RMSSD, freq. answer std., answer std., question trend, and all trends, which exhibited the largest effect on all three models.

Our evaluation method automatically evaluates all statements of meeting participants by using the evaluation indicators mentioned above. Let the weighted average value of the value of each indicator be the evaluation of one statement, and let the sum of the evaluation values of all statements of a participant be the numerical value expressing that participant's speaking ability in discussions at a meeting. By looking at the changes for each discussion in each meeting, participants will be able to judge whether their discussion skills are rising or stagnating.

5.7.2 Virtual Reality Presentation Training System

It is special to perform presentations in front of a large audience, and training such as rehearsal for that is important. However, it is difficult to prepare such a situation in advance. Virtual reality (VR) provides a means for that. In a VR space, you can display a virtual audience who stands on a big stage such as an auditorium, and reacts according to the content of the presentation and the way of presenter's speaking.

We developed a VR presentation training system as shown in Fig. 5.22. In this system, based on the acoustic features of the speaker's voice introduced in Chap. 4 and the characteristics analyzing the slide used for the presentation, we control the reaction of the virtual audience. Also, by tracking the movement of the body with the VR system, we evaluate gaze and gesture and score the presentation. Actions such as responding quickly to the reaction of the audience, changing the way of speaking, changing the display slide, and changing the topic flexibly are also subject to evaluation.

Such a VR training system operates in our smart learning environment. The system includes contents of VR environments that precisely reproduce institutions such as actual lecture halls and auditoriums, 3D models, and action scripts that visualize the appearance and reaction of audiences three-dimensionally. A physical device is

5.7 Future Features of Smart Learning Environments

Fig. 5.22 VR presentation training system (The figure above shows the presenter's 3D avatar and the figure below shows the 3D avatars of the audience)

a headset called head-mounted display (HMD) and a device called a hand tracker (shown in Fig. 5.23) attached to the hand. The HMD incorporates a mechanism for precisely measuring the position and motion of the head and a mechanism for three-dimensionally displaying images and sounds. The hand trackers not only measure the position and orientation of both hands but also the angle at which the finger is bent can be detected in real time. This allows the training system to recognize eye contact and gestures with the audience.

The configuration of hand tracker is as follows.

The bend sensor measuring the bending fits each finger by the cloth glove. Then, this value is transmitted to the PC using Bluetooth low energy (BLE). At that time, using the inertial measurement unit (IMU) attached to the back of the hand, the values of acceleration/angular velocity of XYZ-axes are also measured and transmitted at the same time.

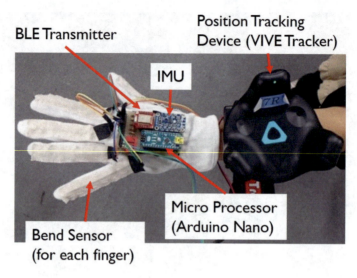

Fig. 5.23 Hand tracker

In addition, the hand tracker is equipped with a position tracking device called the VIVE tracker provided by HTC (a 3D position and orientation tracking device using infrared markers installed in the environment). Therefore, it is possible to know the three-dimensional pose of hands in the tracking area of the VIVE tracker.

A noteworthy point of this system is the feedback method of audience reactions and evaluation results by artificial intelligence. The audience in the VR system is all expressed by 3D animation imitating humans as shown in Fig. 5.22. In the simulation of the reaction, we measure actual human data and use what is modeled by machine learning. The data obtained from the human audiences in the presentation (wearing the HMD and the hand trackers as well as the presenter) are the position and direction of the head and body, the myoelectric potential, the heart rate. They are acquired by dedicated sensors in addition to the VR system. These data are collected by operating a meeting system using VR described below. The audience's model is accompanied by parameters that adjust the degree and timing of reaction, and ingenuity is given to prevent all people from always showing similar reactions to the same stimulus.

Feedback on the presentation is displayed as a reaction of the audience during the presentation. Also, after the presentation, it is displayed as a score for several features mentioned above and can be browsed at an arbitrary timing together with a history of past presentations. The history of the presentation in the VR system is in chronological order of 3D animations with sounds of their own avatar presenting using slides and other materials.

Like the evaluation indicators of the discussion mentioned in Chap. 4, the evaluation indicators of the presentation are based on acoustic features such as voice size, intonation, and speech speed and those based on eye contact between presenter and audiences. In addition, the simplicity of the content of presentation materials

5.7 Future Features of Smart Learning Environments

such as slides, the balance of charts (proportion of the whole document), and the relevance of explanation by voice and the region designated by the pointer are taken into consideration.

In the VR presentation system, the pointer is operated by pointing with a hand tracker, and the pointing part is highlighted. Although it operates differently from a normal laser pointer, it can realize almost the same function as the pointing system using the discussion commander in the discussion mining system described in Chap. 2.

Of course, this VR presentation system can be used not only for training but also for meetings such as laboratory seminars. In that case, the audience is not an AI but an actual human avatar, and participation from a remote place is possible. In that case, human avatar can input gesture using hand trackers, and head movement is also reproduced on VR using HMD.

Also, in the case of a meeting by VR, as in the case of the face-to-face meeting, the contents can be recorded in detail and used for searching, knowledge discovery, and so on. Since the VR system is equipped with a voice recognition function, it is possible to convert the voice of the speaker into text, and furthermore, by using the pointer function by the hand tracker, in addition to the movement of the head and the hand, it is possible to record pointing and referring actions. In this system, since the meeting participants are wearing the HMD, it is impossible to recognize the facial expression from the image, but it can be estimated by using the myoelectric sensor brought into contact with the skin of the face. Information on gestures and facial expressions of meeting participants are used to construct a machine learning model of reaction of audience AI in the presentation training system.

The machine learning model represents relationships among various data gathered by the presentation training system, the behavior of the audience (raising the face, cranking the neck, laughing, sleeping, hitting the hands, stomping the feet, etc.), and mental conditions of the audience (interested, not interested, pleased, angry, etc.). Based on the relationships, the VR system controls the reaction of audiences. While seeing this reaction, the presenter examines the presentation state (whether it is well accepted by the audience) and considers how to improve the presentation.

We are not conducting subject experiments of this system yet, but we will demonstrate that students' presentation skills will improve by operating this VR training system for a certain period of time.

References

K. Armstrong, Big data: a revolution that will transform how we live, work, and think. Inf. Commun. Soc. **17**(10), 1300–1302 (2014)

Y. Bengio, Learning deep architectures for AI. Found. Trends Mach. Learn. **2**(1), 1–127 (2009)

A.J. Camm, M. Malik, J. Bigger, G. Breithardt, S. Cerutti, R. Cohen, P. Coumel, E. Fallen, H. Kennedy, R. Kleiger, Heart rate variability: standards of measurement, physiological interpretation and clinical use. Eur. Heart J. **17**(3), 354–381 (1996)

L. Cutrone, M. Chang, Kinshuk, Auto-assessor: computerized assessment system for marking student's short-answers automatically, in *2011 IEEE International Conference on Technology for Education (T4E)*, pp. 81–88 (2011)

G.R. Hess, K.W. Tosney, L.H. Liegel, Creating effective poster presentations: AMEE guide no. 40. Med. Teach. **31**(4), 319–321 (2009)

D. Iakovakis, L. Hadjileontiadis, Standing Hypotension Prediction Based on Smartwatch Heart Rate Variability Data: A Novel Approach, in *Proceedings of the 18th International Conference on Human-Computer Interaction with Mobile Devices and Services*, pp. 1109–1112 (ACM, 2016)

D. Kerby, J. Romine, Develop oral presentation skills through accounting curriculum design and course-embedded assessment. J. Educ. Bus. **85**(3), 172–179 (2009)

K. Kurihara, M. Goto, J. Ogata, Y. Matsusaka, T. Igarashi, Presentation Sensei: A Presentation Training System using Speech and Image Processing, in *Proceedings of the 9th International Conference on Multimodal Interfaces, ICMI'07*, pp. 358–365 (ACM, New York, NY, USA, 2007)

A. Luque-Casado, M. Zabala, E. Morales, M. Mateo-March, D. Sanabria, Cognitive performance and heart rate variability: the influence of fitness level. PLoS ONE. https://doi.org/10.1371/journal.pone.0056935 (2013)

S. Ohira, K. Kawanishi, K. Nagao, Assessing motivation and capacity to argue in a gamified seminar setting, in *Proceedings of the Second International Conference on Technological Ecosystems for Enhancing Multiculturality. TEEM'14*, pp. 197–204. ACM, New York, NY, USA (2014)

Y. Omori, K. Ito, S. Nishida, T. Kihira, Study on supporting group discussions by improving discussion skills with ex post evaluation, in *IEEE International Conference on Systems, Man and Cybernetics, 2006. SMC'06.*, vol. 3, pp. 2191–2196 (2006)

T. Pereira, P.R. Almeida, J.P. Cunha, A. Aguiar, Heart rate variability metrics for fine-grained stress level assessment. Comput. Meth. Programs Biomed. **148**, 71–80 (2017)

S. Peng and K. Nagao, Automatic Evaluation of Presenters' Discussion Performance based on Their Heart Rate, in *Proceedings of the 10th International Conference on Computer Supported Education (CSEDU 2018)* (2018)

A.J. Sellen, S. Whittaker, Beyond total capture: a constructive critique of lifelogging. Commun. ACM **53**(5), 70–77 (2010)

H. Trinh, K. Yatani, D. Edge, PitchPerfect: Integrated rehearsal environment for structured presentation preparation, in *Proceedings of the 32nd Annual ACM Conference on Human Factors in Computing Systems, CHI'14*, pp. 1571–1580 (ACM, New York, NY, USA, 2014)

N. Wanas, M. El-Saban, H. Ashour, W. Ammar, Automatic scoring of online discussion posts, in *Proceedings of the 2Nd ACM Workshop on Information Credibility on the Web. WICOW '08*, pp. 19–26 (ACM, New York, NY, USA, 2008)

X. Wang, X. Ding, S. Su, Z. Li, H. Riese, J.F. Thayer, F. Treiber, H. Snieder, Genetic influences on heart rate variability at rest and during stress. Psychophysiology **46**(3), 458–465 (2009)

Chapter 6
Symbiosis between Humans and Artificial Intelligence

Abstract In the last chapter, I will talk about how artificial intelligence develops and supports humans in the future, and how humans and machines harmoniously cooperate to create a new society. Symbiosis between humans and artificial intelligence can be said to be a relationship that enhances each other's abilities. In order to establish such a relationship, one's learning needs to have a positive influence on the other's learning. The current mainstream is to advance artificial intelligence by machine learning based on human-made data. However, if artificial intelligence can properly support human learning and human beings can generate useful data for artificial intelligence as a byproduct of humans' main activities, positive circulation will be established between human and artificial intelligence learning. At this time, artificial intelligence needs to understand the situation of human beings, support learning, and actively collect data necessary for their learning. Such artificial intelligence should be an autonomous existence. Indications of autonomous artificial intelligence have already appeared. One example would be a smart speaker like Google Home. It usually listens to the human calls silently, and when it is called it responds to it. However, if a universal pattern is found in human behavior (happen at the same time every day at the same time, sleep at the same time every day, when we wake up in the morning, first check the weather, etc.), artificial intelligence could also prefetch human demands and provide information. We need to think about evils when such autonomous artificial intelligence becomes common. It begins with a leak of privacy, which in turn involves human control and management. To ensure that artificial intelligence does not become a disadvantage to humans, we should establish a symbiotic relationship with artificial intelligence as soon as possible. In this chapter, I will also discuss that point.

Keywords Autonomous artificial intelligence · Intelligent agent · Chatter bot · Smart speaker · Singularity · Human–computer symbiosis

© Springer Nature Singapore Pte Ltd. 2019
K. Nagao, *Artificial Intelligence Accelerates Human Learning*,
https://doi.org/10.1007/978-981-13-6175-3_6

135

6.1 Augmentation of Human Ability by AI

As mentioned in the previous chapter, humans will properly extend their abilities, and machines will add new functions to it. Machines will contribute to supplementing human memory and judgment because they are good at search based on large amounts of data and prediction based on them.

In the explanation so far, artificial intelligence is rather playing a backseat role, and it does not claim any self-assertion, but we can think of a way of artificial intelligence that is not so. This includes cases where the machine moves away from humans and acts autonomously, for example, a self-driving car that a human driver does not ride. However, strictly human control is supposed to be applied and it will be possible to give instructions to artificial intelligence by remote control. The machine acts autonomously if human instructions do not make it in time or it cannot be instructed by human ability in the first place.

Such autonomous artificial intelligence may become a dangerous existence. However, it is not a rebellion of artificial intelligence, as is told in science fiction novels. In my opinion, the rebellion of artificial intelligence (an event that looks like it) is mostly caused by human error (or negligence); I think that it can prevent recurrence. In other words, I think that artificial intelligence does not directly harm humans by their own judgment (prediction acquired from data). For example, it cannot be said that there are no cases where a person dies due to judgment of artificial intelligence, such as a death by accident from self-driving car, but it is not a case that artificial intelligence acts to killing humans, I think that it is caused by unfortunate coincidence because it is impossible to predict everything that could happen.

The real danger lies in losing one's drive for self-improvement due to reliance on artificial intelligence. Many people often point to Japan's adoption of "Yutori education" (relaxed education) as being a cause of the declining scholastic ability in students. However, I believe that the main cause lies in IT devices and applications typified by smartphones giving children a groundless sense of knowing everything, which leads to their abandoning efforts to learn by themselves. For example, a modern child might wonder what is the purpose of finding out something on my own, when I can just search for anything I am interested in and find it easily. As a result, we now see that the number of young people with low adaptability and lacking fundamental ability has increased.

Anyway, what I would like to claim in this book is that humans cannot adapt to the new society unless efforts continue in the future, and that artificial intelligence has the ability to properly support them. There, we can find out the way of new symbiosis of man and machine.

I think that what we should consider to expand human ability is what something humans cannot replace with machines. Things that have been very difficult to automate until now are gradually becoming feasible by collecting and analyzing large amounts of data that measured the phenomena. It is the stance of this manual that it is not appropriate to let a machine's intellectual creation activities act on behalf of machines, but still I think that someday humans will leave many parts of intel-

6.1 Augmentation of Human Ability by AI

lectual activities to artificial intelligence, for example, writing a novel, writing a musical script, writing a drama script, designing a product, doing medical diagnosis, determining a company's management strategy, and so on.

Of course, it may not be possible to produce very good results at the beginning, but by constantly preparing the evaluation mechanism, always providing feedback and updating the machine learning model, it is possible to achieve reasonable accuracy. I think that scientific basis will be gained about whether the features humans have focused on (or the lessons learned in the past) were appropriate.

However, humans still should continue intellectual creative activities in the future. That is because the current artificial intelligence is a mechanism to learn from the data. I believe that humans will always be able to create new data from creative activities, as long as they strive diligently to do so. However, as long as artificial intelligence is always based on historical data, it seems that what they produce will be a repetition of similar things. Humans always seek new things. Therefore, we can create something new. There is also the term "warm innovation." Therefore, depending on the field, there is a possibility that new data can be created by analyzing past data well. However, I think that creative activities have more than what is obtained from historical data.

The machine learning model and the human brain structure are actually quite different. The neural network that machine learning (especially deep learning) uses is originally made with reference to the human brain, but there are few differences between what we know about brain mechanisms at that the time and what we currently know, there is a gap. In terms of the fact that the brain consists of a network of units called neurons, it is similar to the machine learning model, but its input/output relationship is very different.

The relationship between layers of the neural network in machine learning has no particular meaning. It only specifies a nonlinear function to compute parameters. It is said that neurons of the human brain always output forecast values (Hawkins and Blakeslee 2005). This means that upon receiving a signal in one state, it will output a signal in a later state in time. For example, when listening to a song, the neuron is predicting the next note. Also, when getting off the stairs, neurons are predicting when the foot will touch the next area. When throwing the ball to someone else, the neuron is predicting that the ball will approach the other person. Of course, the neuron sees only one element when expressing a phenomenon in a distributed representation (vector expression), but still predicts the next state of the element. In addition, context information is used for prediction, but its description is somewhat complicated, so I omit it here.

Incidentally, there is an illusion of the eye to a phenomenon famous for human vision (the same line of length looks like a different length, a parallel line looks bent, etc.). This phenomenon is recognized by the human brain as always making forecasts, as initially recognized as predicted, but it is found that the prediction is incorrect while doing some operation, such as drawing an extension line. It seems to be seen correctly. It is probably due to such a brain's work that you can see something that should not exist, or you can hear sounds that cannot be heard (which

138   6 Symbiosis between Humans and Artificial Intelligence

cannot be detected with a microphone). Naturally, this is not all bad; due to the brain's forecasting mechanism, recognition has been accelerated and balanced with behavior.

Anyway, what I want to say is that artificial intelligence and the human brain structure are currently different. So, the way creativity comes up is different. We cannot fully automate our creative activities with machines so far, but we can make good use of each other's characteristics to make it more sophisticated and efficient. This is considered as " augmentation of human ability" in this book.

Augmentation of human ability on the premise of artificial intelligence is to make human and machines work properly, and they are closely related to each other so that humans can achieve greater results than by themselves. Alternatively, it solves problems that cannot be solved by humans alone. Problems that are difficult to solve include not only that it is difficult to solve but also problems that cannot be solved within a limited time, even if you can afford it if you have time to spare. For example, it would be a problem for me to decide what roads I will walk within a few seconds.

Humans must properly communicate their intention (goal) and current state of their actions to the machine. One of Google's founders who swept the world with search engines said that "searching is an act of communicating what people want to do or things they are interested into machines." When we live in this era, when we have something to worry about, we immediately started searching the net (inputting keywords into Web search engines) using PCs and smartphones. I do not think that is a bad thing, but it is not a creative activity if it is completed by the act of investigating. I hope to make a habit of utilizing the retrieved results for subsequent activities and communicating the results to the machine again.

Even if the act of communicating your intention is a daily search, we still must communicate our current state to the machine. Even if we understand human intentions, if we do not know the state of the human, the machine cannot predict what people should do. For example, when considering directions on a map, the destination alone is insufficient and it is necessary to know the current position (or departure point). There are various ways the machine knows the current state of humans. One way to do this is pattern recognition such as image recognition and speech recognition, as explained in Chap. 3. In addition, there is a way to estimate the current state using past history. Pattern recognition as well as estimation from history is current areas of artificial intelligence, that is, because machine learning works effectively.

Especially pattern recognition processes information of the real world (our living physical world). For that purpose, we use a device called a sensor (a camera and a microphone). And, as a matter of course, information that can be used (input) becomes more diverse as more sensors are added. The sensors we are currently paying attention to are three-dimensional sensors and biosensors (sensors that detect biological signals). Three-dimensional sensors are used by machines to learn more about the physical space around humans. In addition, the biosensor is used by the machine to know the internal state of humans (tired, feeling good, etc.).

Also, machines will be able to recognize what people are watching, what they are listening to, and even voices. By processing such a wide variety of data (called

6.1 Augmentation of Human Ability by AI 139

multimodal data), the machine can better understand the state of humans. And when you know the human intention and state, artificial intelligence will predict human behavior and give us insight.

Whether humans adhere to that advice or not, perhaps humans can do more than now. It is because humans are not good at objectively monitoring their state (especially internal state), as they are often conscious only after being told by others. In general, the humans brain does not work well when he or she just woke up. Likewise, there are things that humans can do well and things that cannot be done properly depending on their condition. By choosing the most successful actions in the current state, humans will be much better at using time. That means exactly that human capabilities will be expanded.

Just as artificial intelligence is useful for improving specific skills such as discussions, the skills of various creative activities will be improved appropriately with the support of artificial intelligence. The author cannot confirm the effectiveness of all abilities, but we feel that we probably have a rough idea of the things that humans must do in the future. Humans will continue their creative activities in the future, continue to make new ones, and to successfully use artificial intelligence as a tool, so that humans devise to give as much data as possible to artificial intelligence.

6.2 Intelligent Agents

An artificial intelligence system actively acting on humans is called an intelligent agent. Intelligent agents can act actively on their own goals. Therefore, for humans, I think that it can be felt as a system with some personality. However, the personality is not naturally occurring; it is just made to imitate the behavior of humans, and it is perhaps better to call it a pseudo-personality. Since it is not clear how to generate self-consciousness, it is only possible to imitate it. Nevertheless, there is great significance in humans recognizing an agent as being a person and treating them as such.

This is because humans can interact with it in a way they do not normally interact with machines. That is by way of speaking. In Chap. 3, I explained the speech recognition function installed in the meeting recorder, but in this system, artificial intelligence is listening to what humans are talking about, for the purpose of preparing the minutes. It is, of course, an important function, but unlike an intelligent agent, it was not something that actively encouraged humans. The agent listens to human words in order to return some response to humans. In other words, it is very appreciated by the humans to talk to the agent.

A typical example of such a system would be called a chatter bot. This works on a system of sending short messages like Twitter in a textual interactive system that exists on the net. In that case, it interrupts when people are chatting with each other and delivers messages to unspecified people. Some are running with the intention of deceiving humans, and others have data collection purpose to understand human language. However, as of now, there seems to be no chatter bot that is wise enough to successfully deceive many people.

There is a system called ELIZA which can be called the ancestor of the chatter bot (Weizenbaum 1966). ELIZA is a system created by Joseph Weizenbaum in 1966 and interacts with humans like a counselor. For example, if a human says, "my head hurts," ELIZA returns with "Why do you say your head hurts?". And if he or she says, "my mother hates me," ELIZA returns "does anyone in your family and others dislikes you?". ELIZA seems to mimic the reaction of a psychotherapist in a psychiatric interview. Weizenbaum chose psychotherapy to avoid the problem of having the system data on real-world knowledge. Although the situation of psychotherapy is a dialogue between humans, there is a characteristic that almost no knowledge about the content of the dialogue is required. For example, even if a human asked "who is your favorite composer?", ELIZA can return "who is your favorite composer?" or "is that question important?". It seems to be because the dialogue system does not need knowledge about the composer.

Although a chatter bot is a good partner to kill time chatting with, it is not enough to expand human ability, so there is need for more specific intelligence. As an example of such a system, there is a spoken dialogue robot that the authors previously created (Nagao 2002). As shown in Fig. 6.1, this robot has a face and responds variously according to human speech. In some cases, images and images are displayed on the nearby display, and they refer to it and explain by voice. This was made in 2000, at that time, the mechanism of machine learning was not implemented. Therefore, the types of questions that can be answered were quite limited. In other words, the robot could not learn words and knowledge while talking to humans.

Meanwhile, recently, there is a smart speaker in an agent-like system which is attracting much attention. Starting with Amazon Echo, Amazon launched in 2014, Google released Google Home in 2017, and Apple released HomePod in 2018. Figure 6.2 shows the appearance of each product.

If it is placed somewhere in the room, it will retrieve information and manipulate appliances such as lighting and air conditioning according to human calls. Also, since it is a system for playing music, it will play when you request a favorite song. Both the smartphone and the PC operated by touching directly, the distance from the machine becomes a more freeing by being able to operate it by voice. Of course, there was a mechanism to remotely control home appliances, but it was necessary to use a machine (remote control) for that. Since voice can be sent without using tools, there are some definite merits. For example, it is not necessary to remember the procedure for operating the machine, and you can pay attention only to the operation. In other words, you can directly input requests like "I want to know..." and "what is...?".

This is something the researchers of speech recognition have been saying for quite a while but finally it is becoming a reality. For agents, it is a very important function to interact with humans, so it is inevitable that agents become more sophisticated with the development of spoken dialogue technology.

By the way, the biggest difference between the above-spoken dialogue robot and the recent smart speakers is not just whether there is a face. The real difference lies in whether the intelligence system is in the machine before your eyes or whether it exists somewhere else. The mechanism of doing a specific service on the Internet is called the cloud, but for smart speakers, the body of intelligence is in the cloud.

6.2 Intelligent Agents

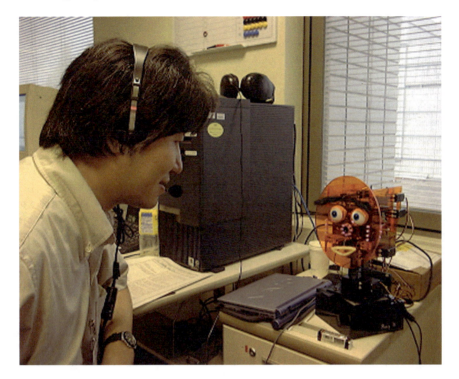

Fig. 6.1 Spoken dialogue robot

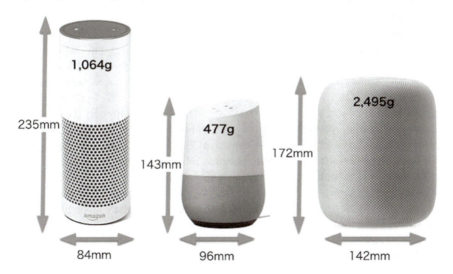

Fig. 6.2 Smart speakers (From Left, Amazon Echo, Google Home, Apple HomePod)

The advantage of having intelligence implemented in the cloud is that it makes machine learning easier. As I have said so far, for machine learning, in most cases, it is advantageous to have as much data as possible. Since the cloud can connect with various machines, the incoming data will be much more than a single machine.

There is not an appropriate place to realize artificial intelligence as much as the cloud, but there are some evils caused by it. That is, the human's live voice (in some cases, including the surrounding video data) is sent to the cloud. For learning, it is better to input the signal of the real world itself, but of course it may be a problem.

Individuals have privacy. Privacy is a complex concept, in short, it can be said that freedom not to disclose personal life information about individuals and freedom that does not receive interference of others against individual activities. However, unintentionally information on privacy may be leaked to the outside. If it is due to a person's mistake, such as what happened due to incorrectly setting the disclosure range of information posted to SNS, it likely cannot be helped. However, in many cases, it is not always the responsibility of the person himself/herself that the problem occurs. Many of the issues come from the cloud.

The problem with smart speakers is that some of the conversations you usually talk about are sent to the cloud. It is used as learning data, but it can be used for other purposes as well. For example, it is a clue to estimate the user's daily behavior and living standards. It may be used as evidence to identify a criminal. In fact, there was a case where Amazon requested that the American court submit data of a certain user's voice obtained by Amazon Echo as evidence (although Amazon refused the request).

This problem is a new problem born with the advancement of technology, and it seems that problems similar to this will occur from now on. In 2013, Edward Snowden accused the US government of collecting personal information gathered. It is a so-called "Snowden Incident". In this way, thinking that collecting and analyzing personal information for the purpose of maintaining public security is a sufficient story, irrespective of its compliance. I do not use SNS and do not unnecessarily disclose personal information. Still, it will be difficult to prevent things like getting involved in some kind of trouble and unfairly exposing personal information. By the way, concerning the relationship between privacy and technology, Ann Cavoukian advocates the concept of Privacy by Design (Cavoukian 2011). She insists that in designing and operating an information system, personal information should be controlled by individuals (such as limiting the scope of disclosure and use).

When using the cloud, there is a possible way to do something and send data with care so as not to leak personal information. In that case, the mechanism of intelligence is distributed among machines and clouds that users directly use. In other words, it means that machines used by individuals also have some intelligence. Such a way of thinking is called edge intelligence. Edge refers to the end of the network. For that, we need to be able to use machine learning to a certain extent even with a small amount of data. In this case, the mechanism of deep learning may not be necessarily effective, but logistic regression analysis and active learning such as explained in Chap. 2 will function effectively.

6.2 Intelligent Agents 143

I believe that by implementing intelligent agents based on the concept of edge intelligence, it is possible to realize a personal assistant system that is sufficiently intellectual and also capable of privacy protection. However, the agent will help the work done by humans, but it will not take over. There is a possibility that the agent will be in charge of some part of your work, but you will not be able to delegate everything. The agent is an autonomous system, but it is not a self-contained system. The agent is designed to be a human support system to the end; it must be such that it operates under the control of humans. It is an essential requirement for agents to coexist with humans.

6.3 Singularity

A phenomenon called the singularity or the technological singularity thought to be a possible outcome of the development of artificial intelligence has been widely discussed (Kurzweil 2005). Current artificial intelligence cannot be said to have the same intelligence as humans, but it is certain that progress is accelerating due to the large concentration of data in the cloud and technologies like deep learning. It is predicted that when artificial intelligence surpasses human intelligence and leads technical advancement it will bring about a rapid and large change that cannot be understood at the speed of human thought. This is a phenomenon called the singularity (to be precise at the time that the phenomenon occurs), but I do not know if this is a correct prediction or not. Also, I think that considering the singularity as a precursor of artificial intelligence against humans is not a very appropriate way of thinking.

I think that it is natural that human life will change by advancing technology, and now I think that we as a society depend heavily on IT, especially the Internet. Also, I think that it is no doubt that artificial intelligence will become more sophisticated, and in certain areas (games such as shogi and go as easy-to-understand examples) will have much better ability than humans. But still, I do not think that artificial intelligence will be an omnipotent existence that surpasses humans in every aspect. Of course, artificial intelligence is obviously advantageous because it is an electronic entity without a physical body. Self-propagation of artificial intelligence is fully considered. However, unless autonomous evolution happens and self-consciousness emerges, it does not seem that it will become threatening humans away from the control of the human who is the designer of artificial intelligence.

If such a thing happened, it means that the design made by humans was wrong. In addition, although we cannot assert about the possibility of autonomous evolution, I think that it is impossible for humans to understand the mechanism, which ultimately makes it impossible to implement. As it happened that the act is properly performed, of course it is possible to repeat similar acts, but I think that such learning of behavior and self-consciousness are not continuous.

If human created design was wrong, it should be modified by humans. In other words, even if artificial intelligence is realized, it is only necessary to realize a mechanism for correcting it and incorporating it in advance. There are people who

144   6 Symbiosis between Humans and Artificial Intelligence

predict that it will become very difficult to stop and redo halfway through learning once learning begins, but I do not think there is such a thing. Humans should become smarter with the evolution of artificial intelligence, because I think that a method of properly stopping artificial intelligence can be implemented for the human who designed it.

There is a prediction that artificial intelligence gets smarter and faster and humans cannot go about it. This prediction seems to ignore the possibility of human efforts, and anyway, I cannot agree with it, regardless of its pros and cons. I think it is inevitable to be overtaken by some people and some abilities. However, I think that artificial intelligence will never be hostile to humans, as artificial intelligence keeps the role of continuing to support the expansion of human capabilities, and as we are carefully managing it.

6.4   Human–AI Symbiosis

J. C. R. Licklider published a paper named Man–Computer Symbiosis in 1960 (Licklider 1960). As written in the paper, symbiosis is defined as "living together in intimate association, or even close union, of two dissimilar organisms." In this paper, it is stated as follows. "The hope is that, in not too many years, human brains and computing machines will be coupled together very tightly, and that the resulting partnership will think as no human brain has ever thought and process data in a way not approached by the information-handling machines we know today."

This can be thought of as connecting the human brain directly to the computer. However, what I call "symbiosis of humans and artificial intelligence" is not a form in which humans and computers are physically coupled. Instead, we assume a form that is tied in with both senses and language ability.

The reason for assuming in this manner is related to a nature that I call separability. Separability is a property of whether the work is still feasible in a situation where humans and machines cooperate and work, if for some reason it allows the human burden to increase even if the machine fails. If a human can execute the work even if the machine stops, it is said he or she can be separated from the machine.

At this point, it would seem that humans are somewhat inseparable from computers with regard to the task of searching. In the case of humans interacting with an agent while working, even if the agent stops to function, I believe that a human would be able to continue working with their own efforts. This is because even if humans rely on machines to shoulder part of their thinking, it is not to say that humans are incapable of thinking on their own without the machine. In the case of working with a machine, we can say that we are relying on the machine, but we do not unilaterally depend on them.

If a human connects his/her computer directly to the human brain, it will be quite difficult to separate the human and the computer. I think that the collaboration between humans and agents should be very similar to the collaboration between people, including the way of communicating that information. Or, sometimes, I think

6.4 Human–AI Symbiosis

that relationship is good so that humans are learners and agents become teachers. At that time, I think that the person who finished learning as much as possible should be able to do the job without agent assistance or guidance.

In terms of educational psychology, there is scaffolding. This is a tool (or someone's support) to raise the learner to a level that can accomplish the task that could not be achieved by one person alone. And when it comes to being able to do a specific job, it also means to do education aiming to be able to be executed even without that support.

I expect that humans can be separated from agents by supporting agents based on the concept of scaffolding. However, agents cannot always be a teacher for humans. In the case that the agent can do better with respect to a specific job than a human, it will be able to support such that it can be understood naturally well. Then, I think whether it will be possible for that person to do as it is, even if any human is not as good as an agent.

Well, in the abovementioned paper, the following things are also mentioned.

"However, many problems that can be thought through in advance are very difficult to think through in advance. They would be easier to solve, and they could be solved faster, through an intuitively guided trial-and-error procedure in which the computer cooperated, turning up flaws in the reasoning or revealing unexpected turns in the solution. Other problems simply cannot be formulated without computing-machine aid. Poincare anticipated the frustration of an important group of would-be computer users when he said, "The question is not, 'What is the answer?' The question is, 'What is the question?'" One of the main aims of man-computer symbiosis is to bring the computing machine effectively into the formulative parts of technical problems."

In other words, in order to solve the problem that we face frequently in reality, the problem is solved earlier by human and computer collaborating with trial and error. Here, trial and error by humans and computers means that the computer repeats the process of verifying the hypotheses considered by humans. This is also "work for computers to solve problems" and "work for humans to understand problems." This situation can be said to be one example where a symbiotic relationship between humans and computers is established.

The symbiotic relationship between humans and artificial intelligence (agents) is similar to this idea. Humans and artificial intelligence should coexist with each other so that they can be separated as much as possible and that humans unilaterally depend on artificial intelligence.

6.5 Agents Embedded in Ordinary Things

I would like to show examples where a human and an agent coexist in everyday life. We developed the system called Ubiquitous Talker in 1995 (Nagao and Rekimoto 1995). This is made with the concept of making everyday things as agents and allowing them to interact with humans. Although the term "ubiquitous" has become less used recently, it is based on the idea of Ubiquitous Computing (Mark Weiser

Fig. 6.3 Ubiquitous talker in use

proposed in 1988), the idea that computers are ubiquitously coordinated to expand the human environment (Weiser 1991). Figure 6.3 shows how this system is used.

In this figure, when a person holds the portable device (currently smartphone) that he has in hand, he can communicate with that one by voice. At the time of making this, there was no concept of the cloud, so the agent's intelligence was implemented on the side of the machine used by the user. Currently, in order to incorporate the mechanism of machine learning (especially deep learning), things will talk to humans while communicating with the cloud.

Recently, the term Internet of Things (IoT) came to be used. This is an idea of implementing the function of connecting with the Internet to all things operating on electricity and linking it with the cloud. One characteristic is that items themselves may not have a mechanism for exchanging information with users. In that case, the user can receive services from IoT by accessing the cloud with a smartphone or PC. I think that the important function that IoT should have is to discover useful information for humans by collecting and analyzing data on the environment surrounding humans rather than directly interacting with humans. Probably, IoT devices will need to work with the cloud for this learning.

One example of IoT is a smart light bulb. This has a function that connects each LED light bulb installed on the ceiling or other locations to the Internet (or an intranet limited to the building). And the users can freely change brightness and color using a smartphone or PC. It is a function that is not yet available in the current smart light bulb, but considering which one is near the light bulb, what the person is doing right now, and what kind of environment is desirable, and cooperating with other surrounding lighting and clouds, light bulbs will actively create a lighting environment suitable for the surrounding people.

The previously smart speaker is also a kind of IoT. If IoT can communicate with humans by voice, it will be possible to realize the Ubiquitous Talker previously proposed by the authors as more practical. It will further step forward the symbiosis of humans and artificial intelligence. For example, if you do not know how to use the tools when you want to do the work, ask the tool or the machine how to use that tool. At that time, if your skill level is converted to data, you can make the tool refer to that data. Then the intelligent agent embedded in that tool will guide you how to use it in an easy-to-understand way. If the guide does not go well, you can tell the agent "I want to know more about it." It is very useful information for agents. It will probably be converted to a teacher signal to update the machine learning model of the agent.

And we can realize the symbiosis of human and artificial intelligence with respect to intellectual activities which are very important to humans such as discussions and presentations which we have repeatedly described in this book. In that case, an agent will appear in the place of discussion and presentation. The agent will evaluate humans' discussions and presentations, and will feed the results back. The agent also contributes to creation of minutes. Also, in some cases, it may make statements that stimulate human creativity. At that time, it can be said that the agent is also participating in intelligent creative activities with humans. Agents will also participate in the gamified discussion mentioned in Chap. 4. In that case, the agent may be able to make better statements after being evaluated by other participants and may eventually become an excellent trainer for discussion beginners one day.

I cannot predict how symbiotic relationships between humans and artificial intelligence will transform with the arrival of technological singularity or the emergence of much higher artificial intelligence than at present. I do not know whether their symbiotic relationships are becoming more sophisticated symbiotic relationships with higher capabilities from each other or whether they are becoming unilateral subordination so that humans are inferior to artificial intelligence. Believing that the former will be realized, I will continue my research in the future.

6.6 Artificial Intelligence Accelerates Human Learning

Finally, I will describe the topic "artificial intelligence accelerates human learning" which is also the title of this book. Specifically, it is about research on an AI trainer which improves student's discussion and presentation skills.

In the story of the previous chapters, AI was not playing the role of guiding humans but was to support humans from behind. This is because it was difficult for humans

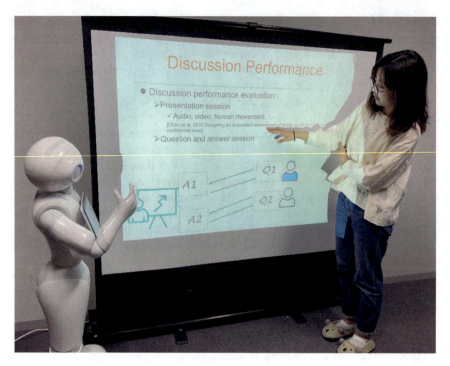

Fig. 6.4 Robot trainer (implemented on pepper by SoftBank Robotics)

to recognize AI as being equal to humans. However, the future AI is not out of sight as a support system. On the contrary, it seems to actively participate with humans as an agent.

The purpose of our research is to intentionally introduce automation with respect to tasks that are concerned about automation and aims to demonstrate with scientific basis that performance will be higher than humans perform. One such task is education, especially dealing with education on discussion and presentation.

Specifically, we are developing a robot trainer (shown in Fig. 6.4) that utilizes various sensors, link with information tools used by students, and bringing about a mechanism to make appropriate suggestions especially for discussion between students and presentations by students. In this task, detailed data analysis will clarify that the skill of students improves when AI analyzes, evaluates, and comments on data rather than teaching by a human trainer.

During the discussion, the robot trainer is always at the side of the student, observing the situation, taking some appropriate timing, and doing some suggestions. For example, as far as the number of utterances is fewer than other participants, it is better to speak more, as to the question better not to return it as a question, better to say the conclusion earlier, and so on.

Likewise, when a student makes a presentation, the robot trainer is always listening nearby, and a comment is made after the presentation. For example, the voice was

6.6 Artificial Intelligence Accelerates Human Learning

difficult to hear, the slide was hard to read, the speaker did not talk properly toward the audience, it was not relaxed, and so on.

The robot trainer can collect and analyze data such as video, voice, person's posture, heart rate, and can decide the contents and timing of the suggestion from various viewpoints. This is similar to the audience AI of the VR presentation training system mentioned at the end of Chap. 5. However, while the audience AI only responds to the human presentation, the robot trainer actively tells humans to promote learning.

Automation in the field of education is not necessarily progressing properly. This is not only technically difficult but also because education is done for humans, and that automation seems to be due to a simple intuition that it does not produce very good effects. The e-learning system and Learning Analytics mentioned in Chap. 1 have increased the number of elements that can be converted to data (Daniel 2017). However, it has not been realized at a manageable level a mechanism that automatically analyzes the data and appropriately reflects it in the guidance.

Therefore, we examine the hypothesis that "the presentation of concrete data on the student's condition and guidance based on it are more likely to be accepted by students than instructed by humans" and realize automation in education. As a result, human learning is accelerated by AI.

The specific verification method will be described below.

Expanding the results so far on the analysis and evaluation of discussion and presentation, we realize a mechanism to improve students' discussion and presentation skills using humanoid robots, and conduct demonstration experiments for laboratory students. For that purpose, we record discussions and presentations, analyze contextual information such as the internal structure of the materials (slides and posters, etc.), visual and auditory scenes of students' activities, participants' mental states, and then evaluate the quality of their activities.

Based on the analysis results of the data, we propose a way to give effective suggestions to humans. In conjunction with the discussion and presentation support systems mentioned in Chap. 5, we develop a robot trainer that evaluates student activities and gives them a suggestion at an appropriate timing. Then, we compare guidance results with human experts and verify the effectiveness of the robot trainer.

In the case of discussion, we will expand the discussion mining system and integrate mechanisms that recognize gestures and expressions of participants in real time. In the case of a slide presentation, software for analyzing the internal structure of the slide and analyzing the browsing behavior of the participant who views the slide is developed. On the other hand, in the case of poster presentations, we plan to develop software to analyze the digital poster creation and operation log, as well as software that will record the presentation and analyze the images and sound captured during recording. Recognition of the position and posture of the surrounding people is also performed by means of depth sensors mounted on the ceiling that are capable of tracking the human body. In addition, wearable devices (e.g., Apple Watch) to measure heart rate are attached to all students, and psychophysiological data are also collected in real time and subject to analysis. These systems are the foundation system for operating the robot trainer.

The robot trainer moves to the venue where the discussion and presentation are being conducted and gives appropriate instructions and advice at appropriate timing near the student who is speaking or presenting. In addition to human facial expression, gaze, and voice, biometric information based on heart rate acquired from wearable devices can also be used, and in collaboration with a discussion/presentation support tool, students' skills can be analyzed and evaluated. Activity records and comments can be reviewed later through the foundation system.

In order to make the evaluation of the robot trainer universal, it is necessary to analyze the record of the discussion and presentation from various viewpoints and establish the evaluation method. This is done by organizing the mechanisms described in Chaps. 4 and 5 and clarifying the evaluation index and its calculation method. For example, in the case of discussion, the following points are considered. The quality and frequency of statements, the attitude toward others' opinions (agree, disagree, disinterest), the size of voice of statements, whether the direction of gaze toward the discussion partner, whether the discussion is diverging or converging, whether it is not diverting the topic, whether it properly responds to questions from others, and the like become the point of view.

Also, in the case of a presentation, it is necessary to decide whether the content such as a slide or a poster is easy for the viewer to understand, whether the theme or result has an impact and gets interest, whether the contents are structured in a well-balanced manner, whether the layout is appropriate, whether the way of presentation (how to watch gaze, voice size, speaking speed, time balance, etc.) is appropriate, whether relaxing during presentation, whether the responder answers the question properly, and so on.

This is based on the guidelines of the American Evaluation Association (http://www.eval.org/). In order to confirm the validity of the automatic evaluation method, students are asked to self-evaluate their own activity records by selecting representative ones. Also, we ask students to evaluate each other's activities mutually. We are planning to compare and analyze these results and the evaluation by the system.

Next, it is necessary to link the evaluation of the discussion/presentation to concrete guidance. Basically, it is the way that the robot trainer is at the side of the student, giving advice so that the evaluation goes up at an appropriate time, or advising not to lower the evaluation. One of the hypotheses of this research is that advice by the machine is more likely to be accepted than advice by humans in the case of advice based on the concrete evaluation index and specific data clarifying the score. If similar failures are repeatedly pointed out, perhaps you would feel uncomfortable if this fact were pointed out by another human. However, in the case of machines, it is not so annoying and we can calmly consider their advice.

Whether a student feels uncomfortable is judged based on psychophysiological information such as a heart rate. In addition, students should be able to view the evaluation of their past activities at any time, so that they can compare their activities with others' activities. Furthermore, we compare the content and evaluation of our activities in chronological order, visualize the degree of growth, and make it possible to discover problems by themselves compared to the growth of others. When that problem occurs again, the robot trainer advises by referring to past guidance so that the problem is solved. If they are aware of the problem, it will be easier to accept instructions from the robot. Applying this mechanism to laboratory students, we confirm that their discussions and presentations are highly evaluated since their skills are improved.

By summarizing the results so far, it is possible to realize a method of automatically evaluating discussion and presentation quality and an effective teaching method for machines. When the student performs a discussion/presentation, the robot trainer evaluates the activities from various viewpoints and makes leading comments. For example, it points out the low number of statements, suggests the timing to speak, and points out the quietness of their voice and encourages the presenter to speak up. Also, it points out if the content of the slide is inappropriate and asks for the presentation of a slide with a more detailed explanation, and so on.

In addition, visualization and presentation of evaluation details and time series data compared with past activities are presented. Next, we compare the degree of the change in the evaluation in the case of direct guidance by human with the case of the robot trainer. In the case of a human trainer, we distinguish between instructing based on the data analyzed by the system and instructing based on the subjective evaluation of the human trainer. And it shows that the degree of improvement of the evaluation is significantly higher in the case of using the robot trainer compared with either case of the human trainer. By doing this, it is possible to empirically demonstrate the case of AI accelerating human learning.

The plan is a 3-year schedule and, as of writing of this book, it is still the first year, so we cannot state the results of this research here unfortunately. I would like to describe the details of this research in the sequel of this book.

References

Cavoukian, in *Privacy by Design: 7 Foundational Principles, Information and Privacy Commissioner of Ontario.* https://www.ipc.on.ca/wp-content/uploads/Resources/7foundationalprinciples.pdf (2011)

B.K. Daniel (Ed.), *Big Data and Learning Analytics in Higher Education: Current Theory and Practice* (Springer International Publishing, 2017)

J. Hawkins, S. Blakeslee, *On Intelligence*, Griffin (2005)

R. Kurzweil, *The Singularity Is Near: When Humans Transcend Biology*, Viking Adult (2005)

J.C.R. Licklider, Man-computer symbiosis. IRE Transac. Human Factors Electron. **HFE-1**, 4–11 (1960)

K. Nagao, Situated Conversation with a Communicative Interface Robot. In *Proceedings of the First International Workshop on Intelligent Media Technology for Communicative Reality* (2002)

K. Nagao, J. Rekimoto, Ubiquitous Talker: Spoken Language Interaction with Real World Objects, in *Proceedings of the Fourteenth International Joint Conference on Artificial Intelligence (IJCAI-95)*, pp. 1284–1290 (1995)

M. Weiser, The computer for the 21st century. Sci. Am. **265**(3)94–104, September 1991

J. Weizenbaum, ELIZA—A computer program for the study of natural language communication between man and machine. Commun. ACM **9**(1), 36–45 (1966)

Printed in the United States
By Bookmasters